SUCCESSFUL MATHEMATICS

NORTH CAROLINA
Common Core Math **1**
Item Bank

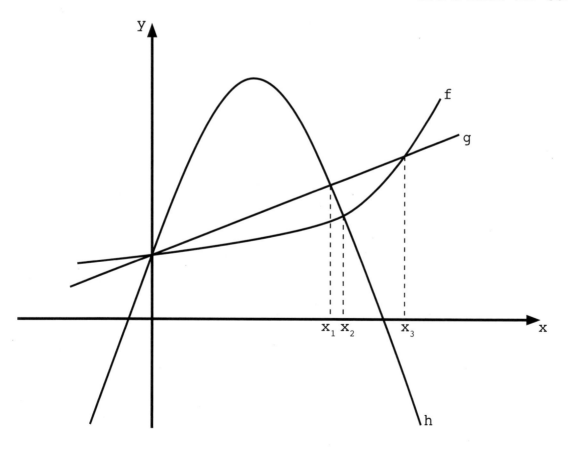

Felix Nagy-Lup

Editor	Felix Nagy-Lup
Cover design	Adrian Mihocas
Text design	Felix Nagy-Lup

VISIT: successfulmathematics.blogspot.com

ORDER INFORMATION:

    E-MAIL: felixnagylup@gmail.com

    PHONE: 910-496-6726

ISBN: 978-0-578-10969-5

I dedicate this book to my precious wife, Emilia, and my precious daughter, Nancy Victoria, whom I thank for all of their support while I was writing it.

PREFACE

This book is written for any student that wishes to study the course: Algebra 1 / Integrated Math 1. It is also for any teacher who is searching for the best usable classroom workbook which contains every one of the new curriculum standards of the Common Core Math Standards.

The book follows the normal teaching of mathematical notation for the student at this level. The exercises in every UNIT are presented in ascending order of difficulty.

In conformity with the new curriculum (Common Core Math Standards), this workbook provides all of the necessary concepts and all of the exercises which the student needs to know in order to be successful in Algebra 1 / Integrated Math 1.

In order to make lesson preparation and homework grading easier for the teacher as they present the new math standards, this workbook presents the 51 standards of the new core in the most logical way, beginning with the easiest and progressing to the more difficult.

Because of the fact that the standards are completely connected to each other, this book presents them one after the other, in a logical and natural order.

The scope of this workbook is the success of the teacher, using a new method to teaching, as well as the success of the student in obtaining more than just a passing grade, but a complete understanding of the material which will lay the groundwork for future success in logical mathematics study.

Contents

Standards	Page

Assessment Quizzes #1-#23

Practice Test

Bibliography

About the author

Recommended PACING GUIDE*

Standards	Number of days
N-RN.1, N-RN.2, A-SSE.2	4
N-Q.1, N-Q.2	2
N-Q.3	1
A-APR.1, A-SSE.2, A-SSE.3	4
A-CED.1, A-CED.4, A-REI.1, A-REI.3	5
A-CED.2, A-REI.10	3
A-REI.5, A-REI.6, A-REI.12, A-CED.3	6
A-REI.11	2
F-IF.1	2
F-IF.2	2
F-IF.3	2
F-IF.4	3
F-IF.5	2
F-IF.6	2
F-IF.7	3
F-IF.8, A-SSE.3	3
F-IF.9	3
F-BF.1	3
F-BF.2	2
F-BF.3	3

Standards	Number of days
F-LE.1	3
F-LE.2	2
F-LE.3	2
F-LE.5, A-SSE.1	2
G-CO.1	1
G-GPE.4, G-GPE.5	5
G-GPE.6, G-GPE.7	3
G-GMD.1, G-GMD.3	2
S-ID.1, S-ID.2, S-ID.3	4
S-ID.5	2
S-ID.6, S-ID.7, S-ID.8, S-ID.9	3

*For block schedule (90 minutes per day)

N-RN.1 *Explain how the definition of the meaning of rational exponents follows from extending the properties of integer exponents to those values, allowing for a notation for radicals in terms of rational exponents.*

N-RN.2 *Rewrite expressions involving radicals and rational exponents using the properties of exponents.*

A-SSE.2 *Use the structure of an expression to identify ways to rewrite it.*

1. Knowing that $x^n = x \cdot x \cdot x \cdot ... \cdot x$ (*n* times), calculate (without using the calculator):
   a. 2^4
   b. 3^2
   c. 5^3
   d. 10^4
   e. 4^3
   f. 2^1
   g. 5^0
   h. 8^2

2. Using the law $x^m \cdot x^n = x^{m+n}$, simplify:
   a. $x^4 \cdot x^3$
   b. $m^2 \cdot m^7$
   c. $a^5 \cdot a^4$
   d. $x^{\frac{1}{3}} \cdot x^{\frac{2}{5}}$
   e. $x^{\frac{1}{6}} \cdot x^{\frac{5}{6}}$
   f. $2^{\frac{4}{5}} \cdot 2^{\frac{6}{5}}$
   g. $3^{\frac{1}{4}} \cdot 3^{\frac{5}{2}}$
   h. $m^{\frac{2}{3}} \cdot m^{\frac{5}{6}}$

3. Using the law $\frac{x^m}{x^n} = x^{m-n}$, simplify:
   a. $\frac{2^8}{2^5}$
   b. $\frac{3^7}{3^9}$

c. $\dfrac{4^{\frac{1}{3}}}{4^{\frac{5}{6}}}$

d. $\dfrac{x^{\frac{5}{6}}}{x^{\frac{4}{3}}}$

e. $\dfrac{5^{\frac{2}{3}}}{5^{\frac{1}{6}}}$

f. $\dfrac{a^{\frac{3}{10}}}{a^{\frac{2}{5}}}$

g. $\dfrac{2^{\frac{3}{4}}}{2^{\frac{3}{4}}}$

4. Using the law $(x^m)^n = x^{m \cdot n}$, simplify:

   a. $(2^3)^4$

   b. $(x^5)^3$

   c. $\left(x^{\frac{2}{3}}\right)^{\frac{6}{5}}$

   d. $(m^2)^5$

   e. $(b^7)^4$

   f. $\left(3^{\frac{1}{2}}\right)^4$

   g. $\left(x^{\frac{3}{4}}\right)^{\frac{4}{9}}$

   h. $(2^6)^{\frac{1}{3}}$

5. Using the laws $(x \cdot y)^n = x^n \cdot y^n$ or $\left(\dfrac{x}{y}\right)^n = \dfrac{x^n}{y^n}$, simplify:

   a. $(3 \cdot x)^2$

   b. $(5 \cdot y^2)^3$

   c. $(4xy^2)^4$

   d. $(x^2 y^3 z^5)^6$

   e. $\left(\dfrac{2x}{y^5}\right)^2$

   f. $\left(\dfrac{x^3}{y^5}\right)^3$

g. $\left(\dfrac{a \cdot b}{c \cdot d}\right)^2$

h. $\left(\dfrac{m^2 \cdot n^5}{p^4}\right)^3$

i. $(3x^5y^2z^4)^3$

j. $\left(x^{\frac{2}{3}}y^{\frac{3}{5}}z^4\right)^{15}$

6. Write the following numbers as products of powers of prime numbers:

    a. 6   **b.** 8   **c.** 10   **d.** 16   **e.** 12   **f.** 18

    g. 28   **h.** 32   **i.** 36   **j.** 40   **k.** 45   **l.** 48

    m. 64   **n.** 72   **o.** 108   **p.** 216   **q.** 324   **r.** 100

7. What is the simplified form of each expression:

    a. $\sqrt[3]{27}$

    b. $\sqrt[5]{32}$

    c. $\sqrt[3]{64}$

    d. $5\sqrt[4]{6} \cdot 2\sqrt[4]{12} \cdot \sqrt[4]{18}$

8. Write each expression in exponential form:

    a. $\sqrt[3]{5^2}$

    b. $12 \cdot \sqrt[3]{x^4}$

    c. $\sqrt[4]{256 \cdot a^8}$

9. Simplify (write the answer in *exponential* form as well as in *root* form):

    a. $\sqrt{7} \cdot 7^{\frac{3}{2}}$

    b. $9^{\frac{1}{3}} \cdot \sqrt[4]{27}$

    c. $8^{\frac{1}{2}} \cdot \sqrt{32}$

    d. $x^{\frac{5}{6}} \cdot x^{\frac{3}{6}} \cdot x^{\frac{4}{6}}$

    e. $9^{\frac{1}{3}} \cdot 27^{\frac{1}{6}}$

    f. $25^2 \cdot \sqrt[4]{125}$

g. $\dfrac{5^{\frac{1}{3}} \cdot 5^{\frac{1}{6}}}{25^{\frac{1}{3}}}$

h. $\dfrac{\left(2^{\frac{1}{3}}\right)^4}{2^{\frac{1}{3}}}$

i. $\dfrac{\sqrt{3} \cdot 3^{\frac{1}{4}}}{\sqrt[4]{3} \cdot \sqrt[3]{9}}$

j. $m^{\frac{1}{4}} \cdot \sqrt[3]{27m^6}$

k. $\sqrt[4]{16a^4b^8} \cdot 3a^2b^5$

l. $\left(x^2 y^{\frac{3}{5}} z^4\right)^{\frac{10}{3}} \cdot \left(x^{\frac{1}{2}} y z^{\frac{2}{3}}\right)^6$

10. What is the simplified form of each expression:

a. $7x^3 \cdot x^7 \cdot 2x$

b. $-5c^3 \cdot 8d^2 \cdot 2c^{-3}$

c. $\left(3a^{\frac{1}{2}} \cdot b^{\frac{1}{5}}\right) \cdot \left(3a^{\frac{1}{4}} \cdot b^{\frac{4}{5}}\right)$

d. $4^{\frac{3}{2}}$

e. $x^m \cdot x^n \cdot x^p$

f. $\left(3t^{\frac{1}{3}} \cdot 7n^{\frac{3}{4}}\right)\left(3t^{\frac{5}{6}} \cdot 7n^{\frac{1}{2}}\right)$

g. $(2j^3 k^5)^{-4} \cdot (k^{-7} j^6)^5$

h. $\left(\dfrac{a^{\frac{1}{4}}}{a^2}\right)^3$

i. $\left(\dfrac{p^{\frac{3}{5}}}{p^2}\right)^{-4}$

j. $(49x)^{\frac{1}{2}}(25x)^{\frac{1}{2}}$

k. $(8x)^{\frac{1}{3}}\left(x^{\frac{2}{3}}\right)$

l. $\left(x^{\frac{1}{2}}\right) \cdot \left(x^{\frac{3}{4}}\right)$

11. The approximate number of Proteins P that a dog needs each day is given by $P = 24 \cdot m^{\frac{3}{4}}$, where m is the dog's mass in kilograms. Find the number of Proteins that a 18 kg dog needs each week.

N-Q.1 *Use units as a way to understand problems and to guide the solution of multi-step problems; choose and interpret units consistently in formulas; choose and interpret the scale and the origin in graphs and data displays.*

N-Q.2 *Define appropriate quantities for the purpose of descriptive modeling.*

1. A fish swims at a rate of 164 feet per hour. Use dimensional analysis to convert this speed to inches per minute.
2. An athlete runs 16 miles in 3 hours. Use dimensional analysis to convert the athlete's speed to feet per second.
3. A monkey can run at a rate of 40 miles per hour. What is the speed in feet per minute?
4. On a map, the distance from New York to Washington DC is 1.75 inches. What is the actual distance, knowing that the scale of the map is 1 inch to 150 miles?
5. The actual distance from Raleigh, North Carolina to Washington DC is 300 miles. What would be the distance on a map that has a scale of 1 cm to 50 miles?
6. A car uses 1 tablespoon of gasoline to drive 90 yards. How many miles can the vehicle travel per gallon? (1 gallon = 256 tablespoons)
7. An airplane flies 25 feet in 3 seconds. What is the airplane's speed in miles per hour?
8. Twenty goats produce 380 pounds of milk. Find the unit rate in pounds per goat.
9. Felix wrote 10 pages of his math homework in 4 hours. What was his writing rate in pages per minute?
10. The maximum speed of a racquet is 950 meters per second. What is this speed in kilometers per hour?
11. The area of a rectangle is at most 40 square inches. The length of the rectangle is 8 inches. What are the possible measurements for the width of the rectangle?
12. Which is a better buy, 8 bagels for $4.24 or 5 bagels for $2.95?
13. Emilia travels from Europe to United States of America. She has Euros, and in order to buy products in USA, she needs to exchange her Euros in dollars. At the airport in Washington DC the exchange rate is: 1 Euro = $1.31 (when you sell) and 1 Euro = $1.37 (when you buy). Emilia trades 800 Euros in dollars. She spends during her vacation $600. With the money left she travels back to Europe. Using the exchange rate at the airport, how many Euros will Emilia get for her unspent dollars?

14. Choose a proper scale and graph the following data:

X	0	100	200	300	400	500	600	700
y	0	3	6	9	12	15	18	21

15. A ball was thrown in the air. The data below shows the height of the ball (in meters) during the 11 seconds it was in the air. Choose a proper scale and graph the following data:

t	0	1	2	3	4	5	6	7	8	9	10	11
h	0	5	12	19	26	30	28	21	18	14	9	0

If the heights would be converted to centimeters, would the scale of the graph be the same? Explain.

16. The table below shows the area of a square for different lengths of its side. The length is given in feet and the area in square feet. Choose a proper scale and graph this data:

Length	0	1	2	3	4	5
Area	0	1	4	9	16	25

If the lengths of the side would be converted to inches, would the scale of the graph be the same?

17. The price of a work of art since 1850 until 2000 is given in the table below:

Year	1850	1875	1900	1925	1950	1975	2000
Value (in thousands of dollars)	235	350	400	470	800	850	925

a. Choose a proper scale and origin and graph this data. Interpret the scale and the origin.

b. If the prices would be converted to millions of dollars (e.g. 235 thousands = 0.235 millions) how would the scale and the origin of the graph modify?

18. What is the actual distance from New York to Los Angeles? Use the ruler to measure the distance from New York to Los Angeles on a map. Find the scale of the map and then calculate the actual distance between the two cities.

19. A usual Formula 1 race lap is 4 km long. A race is completed in 80 laps. One year, the winner's average speed was 230 km/hour. During warming lap runs the average speed was only 130 km/hour. If the race had 10 warming laps, about how long did it take the winner to complete the race?

20. The unit of measure for resistance, R, is called Ohm. We know that $R = \frac{V}{I}$, where V is the potential difference measured across the conductor in Volts and I is the current through a conductor measured in Amperes. What is the unit of Ohm (in terms of Volts and Amperes)?

N-Q.3 *Choose a level of accuracy appropriate to limitations on measurement when reporting quantities.*

1. Two stores have sales on clothing. A same sweater before sale cost $48. The first store offers a $10 off from the original price and the second store offers 25% off from the original price. Which store offers the better deal?
2. In one year, the price of the gasoline rose from $3.20 per gallon to $3.60 per gallon. What was the percent increase?
3. Felix tells Emilia that the distance from Fayetteville to Raleigh is 60 miles. The actual distance is 55 miles. What is the percent error in Felix's estimation?
4. John's height is measured as 54 inches to the nearest inch. What is John's minimum and maximum possible height?
5. The side lengths of a rectangle have been measured to the nearest half of meter as follows: length = 20.5 meters, width = 14.5 meters. What is the greatest possible percent error in finding the area of the rectangle?
6. The dimensions of a gift box are given to the nearest inch: length = 8 inches, width = 5 inches, height = 4 inches. What is the greatest possible percent error in calculating the volume of the gift box?
7. You estimate that your brother's girlfriend is about 16 years old. She is actually 14 years old. What is the percent error in your estimation?
8. Marcos bought a shirt for $45, but the price he paid was 30% off the original price. What was the initial price of the shirt?
9. What would an appropriate level of accuracy be when measuring the length of the shore of a beach? Explain your reasoning.
10. What is the accuracy of a ruler with 8 divisions per inch?
11. What would an appropriate level of accuracy be when studying the number of cell phone users?
12. Emilia the Engineer is designing a truck. She correctly computes that the maximum safe load of the truck being planned will be $100(87 - 50\sqrt{3})$ tons. Felix the Mathematician is the safety supervisor. He is asked to design a sign to tell the drivers how much weight the truck will hold. Felix uses Emilia's expression and uses 1.7 as an approximation for $\sqrt{3}$, and then he creates a sign based on his calculations. The truck is released on the market in 2010. Two weeks later, the truck collapses under a load less than a fourth of the weight shown on Felix's sign. Felix tells the insurance company that he had simply used Emilia's figures. Emilia reports that she has been over and over her figures and can't see how they could be wrong. Write a clear explanation for the insurance company of why the truck collapsed.

A-APR.1 *Understand that polynomials form a system analogous to the integers, namely, they are closed under the operations of addition, subtraction, and multiplication; add, subtract, and multiply polynomials.*

A-SSE.2 *Use the structure of an expression to identify ways to rewrite it.*

A-SSE.3 *Choose and produce an equivalent form of an expression to reveal and explain properties of the quantity represented by the expression.*

> *a. Factor a quadratic expression to reveal the zeros of the function it defines.*

1. Define a *closed* set under a certain operation.
2. Is the set $A = \{-1, 0, 1\}$ *closed* under the *multiplication*? Explain your reasoning.
3. Write each polynomial in *Standard Form*:
   a. $7x + 3 + 5x^2$
   b. $10 + x^2 + 2x^4 + 4x + 5x^3$
   c. $10x - 9 + x^2$
   d. $7 - 3x^2 + 4x$
   e. $5 + 3x$
   f. $2x^3 - x^4 + 5x - 7x^2 + 10$
   g. $7 - 4x$
   h. $2 + 5x - 4x^3 + 8x^2$
   i. $-2x + 5 + 9x^2$
   j. $2 - 9x^5 - 3x^4 + 2x - x^2$
4. If $P = 2x^2 + 4x + 5$ and $Q = 3x^2 + 5x + 10$ are two polynomials, find $P + Q$ and $P - Q$.
5. If $f = 4x^2 + x + 2$ and $g = 5x^2 + 3x + 1$ are two polynomials, find $3 \cdot f$, $2 \cdot g$, $3 \cdot f + 2 \cdot g$, $3 \cdot f - 2 \cdot g$, $f + g$, $f - g$.
6. Simplify:
   a. $x + x$
   b. $x + x + x$
   c. $x + x + x + x$
   d. $x + 2x + x + 3x$
   e. $x - x + x - x$
   f. $x^2 + x^2$
   g. $x^2 + x^2 + x^2$
   h. $2x^2 + x^2 - 4x^2 - x^2$
   i. $3x^2 - x + 5 + 5x^2 - 4x - 9$
   j. $-4x^2 + 3x - 8 - x^2 - 4x + 10$
   k. $a^2 - a^3 + a + 2a^2 - 3a - 4a^3$
   l. $a^2 + a + a^2 + a$
7. Simplify each sum or difference:
   a. $(x^2 + 4x - 5) + (3x^2 - 4x + 9)$
   b. $(-5x^2 + x - 10) + (9x^2 - 3x + 12)$

    c. $(4x^2 - 3x + 9) - (7x^2 + x - 15)$

    d. $(-3x^2 + 7x - 8) - (-5x^2 + x - 3)$

    e. $(x^3 - 4x) - (2x - x^2)$

    f. $(2 + 3x^2 - 5x) + (-7x^2 + 10 - 8x)$

    g. $(9 - x + x^2) - (2x^2 + x - 5)$

    h. $(x^2 + 2x + 3) + (3x^2 + 4x + 10) - (2x^2 + 5x + 9)$

    i. $(4x - 10 - 3x^2) - (5x^2 - 3x + 8) + (x^2 + 2x - 3)$

    j. $(4x^2 - 7) - (2x + 3)$

    k. $(2x^3 - 5x^2 + 8x) - (9x^3 + 3x^2 - 10)$

    l. $-2x^4 + 5x^2 - x^3 + 4x^2 + 3x^4 - 10x + 5 - 9x^2 - 7x^4 - 10x^2 + 3x$

    m. $(-7a^4 + a^2) - (-5a^3 + 8a^2 - a)$

    n. $(-8m^3 + 5m - 4) - (-5m^2 + 2m + 7)$

    o. $(y^3 - 9y^2 - 5) - (7y^3 + 10 - 2y^2)$

    p. $(5z^3 - 8z + 10z^2) + (4z^2 - 3z - 8)$

    q. $(-3b^7 + 4b) - (2b^2 - 10 + b^4)$

    r. $(2c - 3c^4) + (-4c^3 + 9c - 8)$

8. The length of a *rectangle* is $(3x + 7)$ and the width is $(2x + 8)$. What is the perimeter of the rectangle in terms of x?

9. The length of the side of a *square* is $(x^2 + 5x)$. What is the perimeter of the square in terms of x?

10. The perimeter of a *rectangle* is $(2x^2 - 9x + 7)$ and the length of it is $(x^2 + 2x - 3)$. What is the width of the rectangle in terms of x?

11. The perimeter of a *quadrilateral* is $43a + 74$. Three sides have the following lengths: $8a + 3$, $15a + 17$ and $24a + 19$. What is the length of the fourth side in terms of a?

12. Find the sum of the polynomials: $f = -3x^2 + x - 5$ and $g = 4x^2 + x - 10$.

13. Find the difference of the polynomials: $P = 4a^2 - 3a - 9$ and $Q = -3a^2 + a - 10$.

14. Multiply the following monomials:

    a. $x \cdot x$

    b. $2x^3 \cdot 5x^2$

    c. $x \cdot x^2$

    d. $4x^3 \cdot x^2$

    e. $4x \cdot 3x$

    f. $-2x \cdot x^2$

    g. $12x \cdot (-3x^2)$

    h. $-4x^2 \cdot (-5x)$

    i. $2x \cdot x$

    j. $-2x^2y \cdot 3xy^2$

    k. $4xy^3 \cdot 2x^2y$

15. Simplify each product:

    a. $x \cdot (x + 1)$

    b. $x \cdot (x^2 + 2x + 3)$

    c. $x \cdot (x^2 + x)$

    d. $x \cdot (3x^2 + 4x)$

    e. $x \cdot (5x^2 + x + 4)$

    f. $x^2 \cdot (x^2 + x + 1)$

g. $x^2 \cdot (3x^2 - x - 5)$

h. $x^2 \cdot (x - 1)$

i. $x^2 \cdot (4 - x)$

j. $2x^2 \cdot (3 - 2x + x^2)$

k. $4x^2 \cdot (3x^2 - x + 5)$

l. $2x^2 \cdot (4 - x^2 + 3x)$

m. $-3x \cdot (2x^2 - 7 + 2x)$

n. $4xy \cdot (2xy - 1)$

o. $2x \cdot (3x - y)$

p. $-3xy^2 \cdot (2x^2y + 5xy^2)$

q. $5x^2y \cdot (2y - 5x)$

16. Simplify:

a. $x(x + 1) + x(x + 2)$

b. $x \cdot (x - 4) - x \cdot (x + 5) + 2x \cdot (x - 3)$

c. $3x^2(x + 5) - x(5x^2 - 3)$

d. $5a(3a^2 - 4) - a(4a + 1)$

e. $2p(p^2 + 1) - p \cdot (3p) + (4p) \cdot p$

f. $2x \cdot (x^2 - x + 3) - 4 \cdot (x^2 + 3x - 5)$

17. Multiply the polynomials:

a. $(x + 1) \cdot (x + 2)$

b. $(x + 2) \cdot (x + 3)$

c. $(x + 3) \cdot (x + 4)$

d. $(x + 1) \cdot (x + 4)$

e. $(x + 2) \cdot (x + 5)$

f. $(x + 4) \cdot (x - 5)$

g. $(x - 6) \cdot (x + 3)$

h. $(x - 1) \cdot (x - 2)$

i. $(x - 4) \cdot (x + 10)$

j. $(x + 3) \cdot (x - 9)$

k. $(2a - 7)(a + 3)$

l. $(3a + 4)(4a - 2)$

m. $(4a - 3)(5a + 1)$

n. $(2a + 9) \cdot (a - 4)$

o. $(2b - 3) \cdot (-3b + 4)$

p. $(7 - 4b) \cdot (2b - 5)$

q. $(b - 4) \cdot (3b + 9)$

r. $(m - 3) \cdot (7m + 2)$

18. Simplify:

a. $(x + 2)(x + 3) + (x + 4)(x + 1)$

b. $(x + 5)(x + 7) + (x + 3)(x + 4)$

c. $(a - 3)(2a + 5) - (2a + 7)(a - 6)$

d. $(3a + 1)(2a - 5) - (a - 3)(2a + 5)$

e. $(3p + 2)(6p - 5) - (4p + 15)(p + 3)$

f. $(6t - 7)(8t - 1) + (3t - 4)(2t + 1)$

g. $(3 - 5t)(2t - 4) + (t - 3)(3t + 5)$

h. $2m + 3 \cdot (4m - 6) - 5$

i. $5x^2 - 2x \cdot (3x + 4) + 8$

19. Multiply the polynomials:

    a. $(x + 1) \cdot (x^2 + x + 1)$

    b. $(2x + 3) \cdot (x^2 + 2x + 4)$

    c. $(3x + 5) \cdot (4x^2 - x + 2)$

    d. $(4x - 2) \cdot (x^2 - 3x + 5)$

    e. $(x - 3) \cdot (x^2 + 4x - 5)$

    f. $(x - 2) \cdot (2x^2 - x + 4)$

    g. $(2x - 5) \cdot (3x^2 + x - 7)$

    h. $(x + 1) \cdot (x^2 - 2x - 6)$

    i. $(4x - 7) \cdot (2x - 3x^2 + 5)$

    j. $(2b^2 + 1)(b + 5)$

    k. $(m^2 - 9)(3m + 2)$

    l. $(2p^2 + 3)(p + 5p^2)$

    m. $(-c^2 - 2)(c + 3)$

    n. $(3d^2 + 4)(d + 5)$

    o. $(5x - 2)(2x^2 + x - 3) - (3x + 4)(x^2 - x + 5)$

    p. $(x - 1) \cdot (x^2 + x - 2) - (x + 2)(x^2 - 3x + 4)$

    q. $3x(x - 3) - 2x^2(x + 5) - 5(x^2 + 3x - 7)$

    r. $(2x + 1)(7x^2 - x + 5) + (-x + 4)(5x^2 + 2x - 3)$

20. The width of a *rectangle* is $(4x + 1)$ and the length is $(2x + 5)$. What is the area of the rectangle in terms of x?

21. The base of a *triangle* is $(2x + 9)$ and the corresponding height is $(3x - 7)$. What is the area of the triangle in terms of x?

22. The length of the side of a *square* is $(3x + 5)$. What is the area of the square in terms of x?

23. The width of a *rectangle* is $(5x + 2)$ and the length is three times its width. What is the area of the rectangle in terms of x?

24. Simplify:

    a. $(x + 1)^2$

    b. $(x + 2)^2$

    c. $(2x + 3)^2$

    d. $(3x - 5)^2$

    e. $(x^2 + 2x + 3)^2$

    f. $(2x^2 - x + 5)^2$

    g. $(x + 2)^3$

    h. $(2x - 1)^3$

25. Is the set of polynomials closed under the operation of addition? How about subtraction? How about multiplication? Explain your reasoning.

26. Factor, using the Greatest Common Factor:

    a. $4x + 4y + 4z$

    b. $2x - 4y + 10z + 12t$

    c. $8x^3 + 16x^2 - 24x + 40x^4$

    d. $15xy^2 - 25x^2y^2 + 20x^2y^3 - 55x^2y^4$

e. $24a^4b^3 + 42a^3b^3 - 18a^2b$

f. $15a^3b^2 - 20a^4b^2 + 10a^2b^3$

g. $12x^2y + 8x^3y^3 - 24xy + 16x^2y^2$

h. $54x^2 - 36x^4 + 18x^3 - 72x^5$

i. $a(x-3) + b(x-3) - 5(x-3)$

j. $(x+2)(3x-7) - (x+2)(4x-2) + 3(x+2)$

k. $2a(x-1) + 3b(x-1) - 4(x-1)$

27. Factor, using the *"Difference of Two Squares"* formula: $\boldsymbol{a^2 - b^2 = (a-b) \cdot (a+b)}$

    a. $x^2 - 4$

    b. $x^2 - 1$

    c. $a^2 - 9$

    d. $b^2 - 64$

    e. $9x^2 - 4$

    f. $25m^2 - 16$

    g. $4x^2 - 25$

    h. $36p^2 - 1$

    i. $x^4 - y^4$

    j. $b^4 - 16$

    k. $(x+2)^2 - 9$

    l. $(x-3)^2 - 25$

28. Factor the following *Quadratic* expressions:

    a. $x^2 + 5x + 6$

    b. $x^2 + 7x + 12$

    c. $x^2 + 7x + 10$

    d. $x^2 + 9x + 20$

    e. $x^2 - 5x + 6$

    f. $x^2 - 7x + 12$

    g. $x^2 - 9x + 20$

    h. $x^2 + 8x + 15$

    i. $x^2 + 11x + 30$

    j. $x^2 + 15x + 56$

    k. $x^2 - 13x + 42$

    l. $x^2 - 10x + 16$

    m. $x^2 - 9x + 14$

    n. $x^2 - 14x + 48$

    o. $x^2 + 5x + 4$

    p. $x^2 + 7x + 6$

    q. $x^2 + 9x + 8$

    r. $x^2 - 8x + 7$

    s. $x^2 - 4x - 5$

    t. $x^2 - 5x - 14$

    u. $x^2 + 7x - 18$

    v. $x^2 + 8x - 9$

    w. $x^2 - 6x - 27$

29. Factor the following *Quadratic* expressions:
  a. $x^2 + 3x - 4$
  b. $x^2 - 3x - 4$
  c. $x^2 - 4x - 21$
  d. $x^2 + 5x - 24$
  e. $x^2 - 11x - 26$
  f. $x^2 - 10x - 11$
  g. $x^2 - x - 6$
  h. $x^2 + 3x - 10$
  i. $x^2 + 4x - 12$
  j. $x^2 + 5x - 14$

30. Factor the following *Quadratic* expressions:
  a. $3x^2 + 11x + 8$
  b. $4x^2 + 12x - 7$
  c. $4x^2 - 12x - 7$
  d. $2x^2 - x - 3$
  e. $2x^2 - 7x + 6$
  f. $2x^2 + x - 10$
  g. $2x^2 + 7x + 5$
  h. $3x^2 + x - 2$
  i. $2x^2 - x - 3$

31. Factor by grouping terms:
  a. $ax + bx + ay + by$
  b. $5x + 5y - ax - ay$
  c. $x^3 + x^2 - 4x - 4$
  d. $x^3 + x^2 + x + 1$
  e. $x^3 - 3x^2 - 4x + 12$

A-CED.1 Create equations and inequalities in one variable and use them to solve problems. Include equations arising from linear and exponential functions.

A-CED.4 Rearrange formulas to highlight a quantity of interest, using the same reasoning as in solving equations.

A-REI.1 Explain each step in solving a simple equation as following from the equality of numbers asserted at the previous step, starting from the assumption that the original equation has a solution. Construct a viable argument to justify a solution method.

A-REI.3 Solve linear equations and inequalities in one variable, including equations with coefficients represented by letters.

1. Solve each equation:
    a. $x + 3 = 8$
    b. $a - 4 = 15$
    c. $b + 7 = -12$
    d. $m - 9 = -14$
    e. $p + 14 = -13$
    f. $d - 17 = -12$
    g. $7 - x = -23$
    h. $-9 - c = 17$
    i. $-12 - r = -16$
    j. $12 - x = 24$
    k. $35 - x = -27$
    l. $-17 - a = 9$
    m. $-24 - b = -18$
    n. $37 - p = 48$
    o. $2 \cdot x = 14$
    p. $-3 \cdot t = 6$
    q. $-5 \cdot p = -20$
    r. $7 \cdot x = -28$
    s. $-\dfrac{x}{3} = 7$
    t. $-\dfrac{3}{2}x = 6$
    u. $\dfrac{4x}{3} = -8$
    v. $-\dfrac{3}{2}b = 6$

2. Solve each equation:
    a. $2x + 9 = 17$
    b. $-4b - 10 = 18$
    c. $3 - 5a = -12$

    d. $8p - 1 = -9$

    e. $12 = 7 - 5t$

    f. $-14 = 3x - 10$

    g. $-10 = -8 + 2c$

    h. $19 = -10 - 7y$

    i. $43 = 74 + 5k$

    j. $38 = -9x - 17$

    k. $-49 = -14d + 23$

    l. $-10x + 17 = 23$

    m. $\frac{a}{3} - 10 = -12$

    n. $-12 + \frac{3k}{4} = -1$

    o. $17 = 3 - \frac{5}{4}x$

    p. $-10 = \frac{2}{3}p + 17$

    q. $-\frac{1}{5}b + 3 = 7$

3. Solve each equation:

    a. $2x - 5 = -3x + 10$

    b. $-a + 10 = 2a - 8$

    c. $7 - 8b = 10 - b$

    d. $3p + 14 = -24 + 5p$

    e. $-5r + 7 = 8r - 10$

    f. $-14 + 3x = -2x + 6$

    g. $\frac{1}{2}c - 3 = -\frac{2}{3}c + 10$

    h. $5 - \frac{3}{4}t = -12 + \frac{1}{8}t$

    i. $2x - 3 - 7x = 10 - 4x + 6$

    j. $9 - 3x - 10 = 7x - 14$

4. Solve each equation:

    a. $2 \cdot (3x - 5) - 5 \cdot (x + 2) = 4x - 10$

    b. $72 - 3 \cdot (2x - 10) = 10x - 8$

    c. $-3 \cdot (x + 4) + 5 \cdot (3 - 2x) = 2 \cdot (x - 1) - 4 \cdot (3 - 2x)$

    d. $14x - 12 = 3 \cdot (-3x + 5) + 9x - 10$

    e. $4 \cdot (2x - 5) - 3 \cdot (4x - 2) - 10x = 14 + 3 \cdot (4x - 5) - 2 \cdot (7 - x)$

    f. $7 - 3 \cdot (2 - 5a) = 9a + 4 \cdot (a - 3) - 12$

    g. $10b - 4 \cdot (b + 1) = 2 \cdot (3b + 5) - 5 \cdot (2b + 3) - 26$

    h. $7 \cdot (p + 3) - 3p = 5 + 6 \cdot (p + 5) - 4p$

    i. $2 \cdot (5x - 3) + 4x + 3 \cdot (2x - 7) = 5 \cdot (3x + 1) + 8$

    j. $7 - (m - 8) = 4(m + 3)$

5. Solve each equation:

a. $\dfrac{x-1}{3} - \dfrac{2x+1}{5} = x - \dfrac{3}{10}$

b. $\dfrac{2x+3}{5} = \dfrac{x-1}{2}$

c. $\dfrac{x+3}{2} + \dfrac{x}{4} + \dfrac{2x+5}{6} = \dfrac{x+1}{4} - \dfrac{5}{12}$

d. $\dfrac{2x-1}{20} + \dfrac{x+2}{15} - \dfrac{1}{4} = \dfrac{x+1}{12} + \dfrac{7}{12}$

e. $\dfrac{4x+5}{14} + \dfrac{2x-6}{6} = \dfrac{4\cdot(3x-11)}{21} + \dfrac{9}{14}$

f. $\dfrac{3x-5}{4} - \dfrac{x-7}{6} = \dfrac{x-5}{3} + 1$

g. $\dfrac{2x-1}{2} + \dfrac{4x-3}{3} = 2x + 7 - 4\dfrac{1}{6}$

6. Solve the following inequalities:

a. $-2x \leq 6$

b. $4 \leq -2x$

c. $-\dfrac{x}{6} \geq \dfrac{1}{2}$

d. $2x + 11 \geq x + 8$

e. $4x + 10 > 3 \cdot (x + 2)$

f. $3 \cdot (2x - 1) \geq 5 \cdot (x - 1)$

g. $2x - 3 \leq 3x + 1$

h. $2 \cdot (3x + 7) \leq 38$

i. $2 \cdot (x + 5) + 7 < x + 20$

j. $2 \cdot (x - 1) - 1 < x - 4$

k. $4 \cdot (x + 5) - 6 < 2 \cdot (x + 5)$

l. $\dfrac{3x-7}{5} + 2 > \dfrac{x}{3} - 1\dfrac{4}{15}$

m. $\dfrac{2x}{5} - 3 \leq x + \dfrac{1}{2}$

n. $\dfrac{5x-4}{7} - \dfrac{3x+7}{6} < \dfrac{3x-8}{21} - \dfrac{6}{7}$

o. $\dfrac{7x+6}{15} - \dfrac{2x-9}{6} < \dfrac{2x+13}{15} - \dfrac{x+8}{10} + 1\dfrac{1}{3}$

7. Solve each equation for the specified variable:

a. $5p + d \cdot r = r - 7$ for r

b. $\dfrac{m}{n} + 3 = -m + p$ for m

c. $5ab - 3bc = 10$ for c

d. $a = \left(\dfrac{b+c}{4}\right) \cdot d$ for c

e. $N = \dfrac{1}{3}abc + ab$ for b

f. $3d = \dfrac{a-b}{b-c}$ for a

g. $9 \cdot a = b$ for a

h. $2 \cdot b - v = b$ for b

i. $4x - 5y = 8$ for y

j. $ax + b = c$ for x

k. $e \cdot f - 3m = x$ for f

l. $s \cdot t + 5d = 7n$ for t

m. $\dfrac{2x-y}{5} = t$ for y

n. $\dfrac{b-3a}{4} = c$ for b

8. If $y = t \cdot \left(\dfrac{x-u}{2}\right)$ is a given formula:
   a. Find the value of t, knowing that $y = 5$, $x = -3$ and $u = 10$.
   b. Find the value of x, knowing that $y = -2$, $t = 9$ and $u = -7$.
   c. Find the value of u, knowing that $y = 3$, $t = 4$ and $x = -8$.
   d. Solve the formula for t.
   e. Solve the formula for x.
   f. Solve the formula for u.

9. If $d = \dfrac{a-2b}{c}$ is a given equation:
   a. Find the value of a, knowing that $b = -1$, $c = 5$ and $d = -3$.
   b. Find the value of b, knowing that $a = 3$, $c = -2$ and $d = 1$.
   c. Find the value of c, knowing that $a = -6$, $b = 4$ and $d = -5$.
   d. Solve the formula for a.
   e. Solve the formula for b.
   f. Solve the formula for c.

10. If $(2m - 1)x + 3m = m(-4x + 5) + 7x$ is a given equation:
    a. Find the value of x, knowing that $m = 2$.
    b. Find the value of m, knowing that $x = 2$.
    c. Solve the equation for x.
    d. Solve the equation for m.

11. If $(3m + 2)x - 4x = 2m(6x - 1) + 8m$ is a given equation:
    a. Find the value of x, knowing that $m = 3$.
    b. Find the value of m, knowing that $x = -5$.
    c. Solve the equation for x.
    d. Solve the equation for m.

12. If $3x - 5 = -2$, find the value of $-3x + 5$.
13. If $4 = -2x + 8$, find the value of $5x - 9$.
14. The sum of three numbers is 1032. Find the numbers, knowing that the second number is three times the first number and the third number is 60 more than half of the first number.
15. The sum of five consecutive numbers is 175. Find these numbers.
16. A phone company charges $50 per month and 5 cents for each sent text message. Felix cannot spend more than $65 for his phone bill. Create an inequality to find the number of text messages he can send, so he doesn't exceed the $65 limit.

17. The length of a rectangle is $(5x + 8)$ and the width is $(2x + 1)$. What are the possible values for x, if the perimeter of the rectangle is less than 74 yards?

18. The length of a rectangle is $(5n + 7)$ and its width is $(2n − 3)$. The perimeter of the rectangle is twice the sum of its length and its width. What is the value of n knowing that the perimeter of the rectangle is 78 inches?

19. Ken earns a base salary of $800 per month as a salesman. In addition to the salary, he earns $75 per product he sells. If his goal is to earn $2300 per month, create an equation to find how many products does he need to sell in order to reach his goal.

20. A pizza restaurant charges $11.50 for a large cheese pizza. Additional toppings cost $1.30 per topping. Emilia paid $16.70 for her large pizza. Create an equation to find how many toppings did Emilia order for her pizza.

21. The selling price of a refrigerator in a retail store is $500 less than 4 times the wholesale price. If the selling price of the refrigerator is $1200, create an equation to find the wholesale price of the refrigerator.

22. A taxi ride cost $35. The driver charged $7 plus $2 per each mile traveled. Create an equation to find how far did the taxi travel on this trip?

23. Martha has $200. She needs a total of $2,600 to start an account. She earns $75 per day working, of which she saves $60. Create an equation to determine the number of days she needs to work to reach her goal of $2,600.

24. Tania sells paintings in an art gallery. She earns $450 each week plus a commission equal to 3.5% of her sales. This week her goal is to earn at least $1200. Create an equation to find what is the minimum amount of sales she must have this week to achieve her goal.

25. Raphael wants to join a game-of-the-month club. The first club costs $60 to join and $8 per game. The second club costs $11 per game and has no fee to join. Create an equation to find how many games would need to be purchased from each club for the clubs to cost the same.

26. An amusement park charges $12 to enter the park and an additional fee for each time a guest rides a roller coaster. Emilia rode 8 times on a roller coaster. Her total payment was $36. Gabriela rode 7 times on a roller coaster. What was Gabriela's total payment?

27. The sum of two numbers is 88. One of the numbers is 4 more than twice the other number. Determine the two numbers.

28. The sum of three consecutive integers is 75. Determine the three numbers.

29. The sum of three natural numbers is 220. The second number is 30 more than twice the first number and the third number is 10 less than the first number. Create an equation to determine the three numbers.

30. A square and a rectangle have the same perimeters. The length of a side of the square is $(2x + 5)$. The length of the rectangle is $(3x + 4)$ and the width is $(x − 3)$. Create an equation for this context and find the value of x.

A-CED.2 *Create equations in two or more variables to represent relationships between quantities; graph equations on coordinate axes with labels and scales.*

A-REI.10 *Understand that the graph of an equation in two variables is the set of all its solutions plotted in the coordinate plane, often forming a curve (which could be a line).*

1. Graph the equation $y = 2^x$ and find five solutions to this equation. How many solutions does the equation have? How can the solutions be represented on a graph?
2. Graph the equation $y = 4x + 3$ and find five solutions to this equation. How many solutions does the equation have? How can the solutions be represented on a graph?
3. Graph the equation $2x - 3y + 6 = 0$ and find five solutions to this equation. How many solutions does the equation have? How can the solutions be represented on a graph?
4. Graph the equation $x + 2y = 4$ and find five solutions to this equation. How many solutions does the equation have? How can the solutions be represented on a graph?
5. Graph the equation $x = 2y - 8$ and find five solutions to this equation. How many solutions does the equation have? How can the solutions be represented on a graph?
6. Graph the equation $y - 5x = -10$ and find five solutions to this equation. How many solutions does the equation have? How can the solutions be represented on a graph?
7. Graph the equation $y + 3 = 6x$ and find five solutions to this equation. How many solutions does the equation have? How can the solutions be represented on a graph?
8. Graph the equation $x - 6 = 2y$ and find five solutions to this equation. How many solutions does the equation have? How can the solutions be represented on a graph?
9. Graph the equation $y = 5 - 2x$ and find five solutions to this equation. How many solutions does the equation have? How can the solutions be represented on a graph?
10. Graph the equation $x = 10 - 4y$ and find five solutions to this equation. How many solutions does the equation have? How can the solutions be represented on a graph?
11. Graph the equation $4y + x - 12 = 0$ and find five solutions to this equation. How many solutions does the equation have? How can the solutions be represented on a graph?
12. Graph the equation $y = 3 \cdot \left(\frac{1}{2}\right)^x$ and find five solutions to this equation. How many solutions does the equation have? How can the solutions be represented on a graph?
13. Graph the equation $y = -4 \cdot 5^x$ and find five solutions to this equation. How many solutions does the equation have? How can the solutions be represented on a graph?
14. Graph the equation $y = 8 \cdot \left(\frac{1}{3}\right)^x$ and find five solutions to this equation. How many solutions does the equation have? How can the solutions be represented on a graph?
15. Graph the equation $y = -2 \cdot \left(\frac{1}{2}\right)^x$ and find five solutions to this equation. How many solutions does the equation have? How can the solutions be represented on a graph?
16. Felix has $200 to buy books and movie DVD's. Knowing that a book costs $8 and a movie DVD costs $20, create an equation to represent the relationship between these quantities. Graph this equation and find the viable solutions to the problem. What is the

maximum number of books that Felix can buy? What is the maximum number of movie DVD's that Felix can buy?

17. Tatiana has $3,000 in her savings account and she deposits $200 every month. Create an equation in two variables to represent this situation. Graph this equation and find how much money will Tatiana have in her savings account after 3 years.

18. Elis has $36 to buy sodas and ice-creams. Knowing that a soda costs $2 and an ice-cream costs $3, create an equation to represent the relationship between these quantities. Graph this equation and find the viable solutions to the problem. What is the maximum number of sodas that Elis can buy? What is the maximum number of ice-creams that Elis can buy?

A-REI.5 Prove that, given a system of two equations in two variables, replacing one equation by the sum of that equation and a multiple of the other produces a system with the same solutions.

A-REI.6 Solve systems of linear equations exactly and approximately (e.g., with graphs), focusing on pairs of linear equations in two variables.

A-REI.12 Graph the solutions to a linear inequality in two variables as a half- plane (excluding the boundary in the case of a strict inequality), and graph the solution set to a system of linear inequalities in two variables as the intersection of the corresponding half-planes.

A-CED.3 Represent constraints by equations or inequalities, and by systems of equations and/or inequalities, and interpret solutions as viable or non- viable options in a modeling context.

1. Solve the following systems of equations using the *elimination method*:

   a. $\begin{cases} 2x + 3y = 12 \\ 5x - 3y = 2 \end{cases}$

   b. $\begin{cases} x + y = 14 \\ x - y = 26 \end{cases}$

   c. $\begin{cases} 4x - 3y = 18 \\ 4x + 2y = 8 \end{cases}$

   d. $\begin{cases} 3x + 2y = 5 \\ 5x - 3y = 2 \end{cases}$

   e. $\begin{cases} 4x - 2y = 10 \\ 2x + 3y = 1 \end{cases}$

   f. $\begin{cases} x + 3y = -1 \\ 2x - y = 5 \end{cases}$

   g. $\begin{cases} 2x + y = 0 \\ x - 3y = 7 \end{cases}$

   h. $\begin{cases} 4x - 2y = 0 \\ x - y = -1 \end{cases}$

   i. $\begin{cases} 2x + 3y = -3 \\ -x + 2y = 5 \end{cases}$

   j. $\begin{cases} 2x - 3y = 7 \\ 4x - 4y = 4 \end{cases}$

   k. $\begin{cases} 2x + 4y = 5 \\ 4x + 6y = 7 \end{cases}$

   l. $\begin{cases} 4x - 3y = 20 \\ 5x + 2y = 2 \end{cases}$

   m. $\begin{cases} 6x - 2y = -14 \\ 7x + y = -13 \end{cases}$

   n. $\begin{cases} 2x + y = -2 \\ 3x + 2y = -1 \end{cases}$

   o. $\begin{cases} 4x + 2y = -2 \\ 2x + 3y = 9 \end{cases}$

   p. $\begin{cases} 3x + 2y = 1 \\ 4x + 3y = 3 \end{cases}$

27

$$\text{q.} \begin{cases} 5x + 4y = -4 \\ 3x + 5y = 8 \end{cases}$$

$$\text{r.} \begin{cases} 2x + 3y = 11 \\ 4x - 3y = -5 \end{cases}$$

2. Solve the following systems of equations:

$$\text{a.} \begin{cases} 2(x + 3) - 3y = 2 \\ 5x - 3(2y - 1) = -4 \end{cases}$$

$$\text{b.} \begin{cases} 4x - 3y = 13 \\ 2(x - y) + 5y = -7 \end{cases}$$

$$\text{c.} \begin{cases} 4(x + 2) + 3(y + 1) = 15 \\ 5(x + 1) + 4y = 11 \end{cases}$$

$$\text{d.} \begin{cases} 2(x + 1) - 3(y - 1) = -1 \\ 5(x - 2) - 4(y - 3) = 1 \end{cases}$$

$$\text{e.} \begin{cases} 5x + 3(y - 1) = 7 \\ 7(x + 1) + 2y = 10 \end{cases}$$

$$\text{f.} \begin{cases} 2(x + 1) + 3y = 7 \\ 3x + 2(y - 5) = -10 \end{cases}$$

3. Solve the following systems of equations:

$$\text{a.} \begin{cases} 6x - 5.2y = -28 \\ 9x + 2.6y = 10 \end{cases}$$

$$\text{b.} \begin{cases} 2x - \sqrt{3} \cdot y = 5\sqrt{3} \\ 2\sqrt{3} \cdot x + 2y = 30 \end{cases}$$

$$\text{c.} \begin{cases} 2\sqrt{3} \cdot x - 3\sqrt{2} \cdot y = 12 \\ 4\sqrt{3} \cdot x + \sqrt{2} \cdot y = 10 \end{cases}$$

4. Solve the following systems of equations using *graphs* (graph both linear equations and then find the coordinates of the intersection point of the two graphs):

$$\text{a.} \begin{cases} y = 2x - 6 \\ y = -3x + 4 \end{cases}$$

$$\text{b.} \begin{cases} y = 5x - 7 \\ y = -x + 2 \end{cases}$$

$$\text{c.} \begin{cases} y = 12x - 5 \\ y = 1 - 4x \end{cases}$$

$$\text{d.} \begin{cases} y = x + 5 \\ y = x - 2 \end{cases}$$

$$\text{e.} \begin{cases} y = 6 - 5x \\ y = -2x + 9 \end{cases}$$

$$\text{f.} \begin{cases} 2x + 4y = 5 \\ 4x + 6y = 7 \end{cases}$$

$$\text{g.} \begin{cases} 4x - 3y = 20 \\ 5x + 2y = 2 \end{cases}$$

$$\text{h.} \begin{cases} 6x - 2y = -14 \\ 7x + y = -13 \end{cases}$$

i. $\begin{cases} 2x + y = -2 \\ 3x + 2y = -1 \end{cases}$

j. $\begin{cases} 2x + 3y = 11 \\ 4x - 3y = -5 \end{cases}$

k. $\begin{cases} x + y = 12 \\ x - y = -14 \end{cases}$

5. Graph the solution to the following linear inequalities:

    a. $2x + 3y \leq 6$

    b. $-2x + 5y > 20$

    c. $4x - y + 8 \geq 0$

    d. $-x + y + 1 < 0$

    e. $y \geq 3x - 6$

    f. $y < -2x + 8$

    g. $x \leq 2y - 14$

    h. $x > y + 3$

    i. $y - 4x \geq 12$

    j. $-2y + 3x < 18$

    k. $y \geq 10 - 2x$

    l. $x < 3y + 6$

    m. $x - 8 \geq 4y$

    n. $y + 9 < 3x$

    o. $5x + y \leq 5$

    p. $-3x + 4y > 12$

    q. $2x - y + 6 \geq 0$

    r. $x - y + 1 < 0$

    s. $y \geq 2x - 4$

    t. $y < -3x + 6$

    u. $x \leq 4y - 8$

    v. $x > y + 5$

    w. $y - 7x \geq 14$

    x. $-3y + 2x < 18$

    y. $y \geq 8 - 4x$

    z. $x < 2y + 8$

6. Graph the solution to the following systems of linear inequalities:

    a. $\begin{cases} y \leq 2x - 6 \\ y \geq -3x + 4 \end{cases}$

    b. $\begin{cases} y < 5x - 7 \\ y \geq -x + 2 \end{cases}$

c. $\begin{cases} y \geq 12x - 5 \\ y < 1 - 4x \end{cases}$

d. $\begin{cases} y \leq x + 5 \\ y < x - 2 \end{cases}$

e. $\begin{cases} y > 6 - 5x \\ y \geq -2x + 9 \end{cases}$

f. $\begin{cases} 3x + 2y < 6 \\ 5x - 3y \geq 15 \end{cases}$

g. $\begin{cases} 4x - 2y \leq 10 \\ 2x + 3y > 1 \end{cases}$

h. $\begin{cases} x + 3y < -6 \\ 2x - y \geq 5 \end{cases}$

i. $\begin{cases} 2x + y < 0 \\ x - 3y > 7 \end{cases}$

j. $\begin{cases} 2(x + 3) - 3y \geq 2 \\ 5x - 3(2y - 1) \geq -4 \end{cases}$

k. $\begin{cases} 4x - 3y < 13 \\ 2(x - y) + 5y < -7 \end{cases}$

l. $\begin{cases} 4(x + 2) + 3(y + 1) \leq 15 \\ 5(x + 1) + 4y \geq 11 \end{cases}$

m. $\begin{cases} 2x + 3y \leq -6 \\ -2x + 5y > 20 \end{cases}$

n. $\begin{cases} 4x - y + 8 \geq 0 \\ -x + y + 1 < 0 \end{cases}$

o. $\begin{cases} x > 2y - 14 \\ x < y + 3 \end{cases}$

p. $\begin{cases} y - 4x \geq 12 \\ -2y + 3x > 18 \end{cases}$

q. $\begin{cases} y \geq 2x \\ x < 3y \end{cases}$

r. $\begin{cases} x + 3y - 6 \geq 0 \\ x \leq 4 \\ y > -2 \end{cases}$

s. $\begin{cases} y < -2x \\ x > 2 \\ y \leq 5 \end{cases}$

t. $\begin{cases} x > 4y \\ x \leq -2 \\ y \geq 1 \end{cases}$

7. Find two numbers, knowing that their sum is 1240 and one of the numbers is 24 more than three times the other number.

8. The sum of two numbers is 935. Find the numbers, knowing that two thirds of a number is a quarter of the other number.

9. In a building there are 32 apartments with 2 bedrooms and 3 bedrooms. Knowing that there are 86 bedrooms in total, write a system of equations for the context of this problem and find how many 2 bedroom apartments are.

10. In a classroom there are 32 students, boys and girls. If other 5 boys would come in the classroom and 7 girls would leave, then the number of boys would double the number of girls.
    a. Write a system of equations for the context of this problem.
    b. How many boys and girls are there in the classroom?
    c. Is the solution of the system viable for this situation? Explain your reasoning.
11. The sum of two numbers is 623 and the difference of them is 73. Find the numbers.
12. A group of 10 children and 4 adults pay a total of $4500 to take a piano class. A group of 8 children and 6 adults pay a total of $5000 to take the same class.
    a. Write a system of equations for the context of this problem.
    b. What is the total cost for 1 child and 1 adult to take the piano class?
    c. Is the solution of the system viable for this situation? Explain your reasoning.
13. Emilia has in her wallet three times as many 5-dollar bills as 10-dollar bills. She has a total of $200. How many 10-dollar bills does Emilia have in her wallet?
14. A store sold on Monday 250 sodas and 340 ice-creams for a total of $970. The same store sold on Tuesday 280 sodas and 320 ice-creams of the same brand for a total of $980. Knowing that the prices for sodas and ice-creams were the same Monday and Tuesday, what was the cost for one ice-cream?
15. The cost of renting a car for one day includes a flat rental fee and an extra charge for each mile the car is driven while it is rented. The renting price for a car that is driven 200 miles in one day is $195. The renting price for a car that is driven 340 miles in one day is $300. What is the extra charge for each mile the car is driven?
16. A zoo has twice as many birds as reptiles. The total number of birds and reptiles is 120.
    a. Write a system of equations for the context of this problem.
    b. How many reptiles are in the zoo?
    c. Is the solution of the system viable for this situation? Explain your reasoning.
17. A store sold on Monday 80 sodas and 175 ice-creams for a total of $220. The same store sold on Tuesday 60 sodas and 230 ice-creams of the same brand for a total of $323. Knowing that the prices for sodas and ice-creams were the same Monday and Tuesday, what was the cost for one ice-cream? Is the solution of the system viable for this situation? Explain your reasoning.
18. Felix wants to buy books and notebooks from a store. A book costs $8 and a notebook costs $3. Felix has $48 to spend.
    a. Construct a linear inequality for this situation.
    b. What are the possible combinations of books and notebooks that he can buy? Explain your reasoning.
    c. Graph the linear inequality that represents this situation. Interpret the solution and find the viable solutions in this context.

19. At a hotel resort there are two swimming pools that have to be filled with water. One of the pools contains 80 gallons of water and is filled at a constant rate of 15 gallons per hour. The other pool contains 40 gallons of water and is filled at a constant rate of 17 gallons per hour.

    a. Write a system of equations for the context of this problem.

    b. Solve the system to find when will the two swimming pools have the same amount of water and what is that amount of water.

    c. Is the solution of the system viable for this situation? Explain your reasoning.

20. Mrs. Johnson is planning a party for her students. She has 30 students and she wants to buy sodas and ice-creams for all of them. A soda costs $1.25 and an ice-cream costs $1.75. She needs to buy at least 90 products and she has a $120 budget.

    a. Create a system of inequalities that represents this situation.

    b. Graph the system and interpret the solution. Find the viable solution in this context.

21. Felix needs to earn at least $850 a week to be able to pay his mortgage and his bills, but he cannot work more than 60 hours per week. Felix is getting paid $20 per hour for mowing lawns and $10 per hour for washing cars.

    a. Create a system of inequalities that represents this situation.

    b. Graph the system and interpret the solution. Find the viable solution in this context.

22. Emilia wants to buy some shirts and sweaters from an outlet store. A shirt costs $25 and a sweater costs $40. Emilia has a $200 gift card she can use at this store.

    a. Construct a linear inequality for this situation.

    b. What are the possible combinations of shirts and sweaters that she can buy? Explain your reasoning.

    c. Graph the linear inequality that represents this situation. Interpret the solution and find the viable solutions in this context.

23. At a water park there are two swimming pools that have to be filled with water. One of the pools contains 30 gallons of water and is filled at a constant rate of 7 gallons per hour. The other pool contains 20 gallons of water and is filled at a constant rate of 5 gallons per hour.

    a. Write a system of equations for the context of this problem.

    b. Solve the system to find when will the two swimming pools have the same amount of water and what is that amount of water.

    c. Is the solution of the system viable for this situation? Explain your reasoning.

24. Nathan wants to buy CDs and DVDs from a store. A CD costs $3 and a DVD costs $8. Nathan has $30 to spend.

    a. Construct a linear inequality for this situation.

b. What are the possible combinations of CDs and DVDs that he can buy? Explain your reasoning.

c. Graph the linear inequality that represents this situation. Interpret the solution and find the viable solutions in this context.

25. At a hotel resort there are two swimming pools that have to be filled with water. One of the pools contains 10 gallons of water and is filled at a constant rate of 25 gallons per hour. The other pool contains 42 gallons of water and is filled at a constant rate of 17 gallons per hour.

    a. Write a system of equations for the context of this problem.

    b. Solve the system to find when will the two swimming pools have the same amount of water and what is that amount of water.

    c. Is the solution of the system viable for this situation? Explain your reasoning.

26. Mr. Taylor is planning a party for his chess club. He has 20 chess players and he wants to buy sodas and ice-creams for all of them. A soda costs $1.35 and an ice-cream costs $1.55. He needs to buy at least 70 products and he has a $115 budget.

    a. Create a system of inequalities that represents this situation.

    b. Graph the system and interpret the solution. Find the viable solution in this context.

27. Richard needs to earn at least $360 a week to be able to pay his car and his bills, but he cannot work more than 40 hours per week. Richard is getting paid $9 per hour for mowing lawns and $12 per hour for washing cars.

    a. Create a system of inequalities that represents this situation.

    b. Graph the system and interpret the solution. Find the viable solution in this context.

28. Tania wants to buy some shirts and sweaters from an outlet store. A shirt costs $20 and a sweater costs $30. Tania has a $180 gift card she can use at this store.

    a. Construct a linear inequality for this situation.

    b. What are the possible combinations of shirts and sweaters that she can buy? Explain your reasoning.

    c. Graph the linear inequality that represents this situation. Interpret the solution and find the viable solutions in this context.

29. At a water park there are two swimming pools that have to be filled with water. One of the pools contains 26 gallons of water and is filled at a constant rate of 11 gallons per hour. The other pool contains 32 gallons of water and is filled at a constant rate of 14 gallons per hour.

    a. Write a system of equations for the context of this problem.

    b. Solve the system to find when will the two swimming pools have the same amount of water and what is that amount of water.

    c. Is the solution of the system viable for this situation? Explain your reasoning.

30. Christian wants to fence in a rectangular area for a garden. He has only 80 yards of fence. Christian wants the length of the garden to be at least 30 yards and the width to be no more than 16 yards.

    a. Create a system of inequalities that represents this situation.

    b. Graph the system and interpret the solution. Find the viable solutions in this context.

31. The temperature of Cobalt started at 5°C and is increasing by 3°C every hour. The temperature of Zinc started at 9°C and is increasing by 1°C every hour.

    a. Write a system of equations for the context of this problem.

    b. Solve the system to find when will the two chemical substances have the same temperature and what is that same temperature.

    c. Is the solution of the system viable for this situation? Explain your reasoning.

32. Steven wants to fence in a rectangular area for a garden. He has only 50 yards of fence. Steven wants the length of the garden to be at least 15 yards and the width to be no more than 12 yards.

    a. Create a system of inequalities that represents this situation.

    b. Graph the system and interpret the solution. Find the viable solutions in this context.

A-REI.11 *Explain why the x-coordinates of the points where the graphs of the equations y = f(x) and y = g(x) intersect are the solutions of the equation f(x) = g(x); find the solutions approximately, e.g., using technology to graph the functions, make tables of values, or find successive approximations. Include cases where f(x) and/or g(x) are linear and exponential functions.*

1. Solve the equation $2^x = -x + 11$ using graphs:
   a. Graph the equations $y = 2^x$ and $y = -x + 11$ in the same coordinate plane.
   b. In how many points do these graphs intersect? What do these points of intersection represent?
   c. Using a table of values for both equations, find approximately the X-coordinate of the point where the graphs intersect.
   d. Using technology, find exactly the coordinates of the point where the graphs of the equations intersect.
   e. What is the solution of the equation $2^x = -x + 11$? What is the solution of the system of equations: $\begin{cases} y = 2^x \\ y = -x + 11 \end{cases}$? Explain why the X-coordinates of the points where the graphs of the equations $y = 2^x$ and $y = -x + 11$ intersect are the solutions of the equation $2^x = -x + 11$.

2. Solve the equation $5 \cdot 2^x = 4 \cdot \left(\frac{1}{3}\right)^x$ using graphs:

   a. Graph the equations $y = 5 \cdot 2^x$ and $y = 4 \cdot \left(\frac{1}{3}\right)^x$ in the same coordinate plane.
   b. In how many points do these graphs intersect? What do these points of intersection represent?
   c. Using a table of values for both equations, find approximately the X-coordinate of the point where the graphs intersect.
   d. Using technology, find exactly the coordinates of the point where the graphs of the equations intersect.

   e. What is the solution of the equation $5 \cdot 2^x = 4 \cdot \left(\frac{1}{3}\right)^x$? What is the solution of the system of equations: $\begin{cases} y = 5 \cdot 2^x \\ y = 4 \cdot \left(\frac{1}{3}\right)^x \end{cases}$? Explain why the X-coordinates of the points where the graphs of the equations $y = 5 \cdot 2^x$ and $y = 4 \cdot \left(\frac{1}{3}\right)^x$ intersect are the solutions of the equation $5 \cdot 2^x = 4 \cdot \left(\frac{1}{3}\right)^x$.

3. Solve the equation $2x - 6 = -x + 9$ using graphs:
   a. Graph the equations $y = 2x - 6$ and $y = -x + 9$ in the same coordinate plane.
   b. In how many points do these graphs intersect? What do these points of intersection represent?
   c. Using a table of values for both equations, find approximately the X-coordinate of the point where the graphs intersect.

d. Using technology, find exactly the coordinates of the point where the graphs of the equations intersect.

e. What is the solution of the equation $2x - 6 = -x + 9$? What is the solution of the system of equations: $\begin{cases} y = 2x - 6 \\ y = -x + 9 \end{cases}$? Explain why the X-coordinates of the points where the graphs of the equations $y = 2x - 6$ and $y = -x + 9$ intersect are the solutions of the equation $2x - 6 = -x + 9$.

4. Solve the equation $3x + 26 = 5 \cdot \left(\frac{1}{2}\right)^x$ using graphs:

a. Graph the equations $y = 3x + 26$ and $y = 5 \cdot \left(\frac{1}{2}\right)^x$ in the same coordinate plane.

b. In how many points do these graphs intersect? What do these points of intersection represent?

c. Using a table of values for both equations, find approximately the X-coordinate of the point where the graphs intersect.

d. Using technology, find exactly the coordinates of the point where the graphs of the equations intersect.

e. What is the solution of the equation $3x + 26 = 5 \cdot \left(\frac{1}{2}\right)^x$? What is the solution of the system of equations: $\begin{cases} y = 3x + 26 \\ y = 5 \cdot \left(\frac{1}{2}\right)^x \end{cases}$? Explain why the X-coordinates of the points where the graphs of the equations $y = 3x + 26$ and $y = 5 \cdot \left(\frac{1}{2}\right)^x$ intersect are the solutions of the equation $3x + 26 = 5 \cdot \left(\frac{1}{2}\right)^x$.

5. Solve the equation $3 \cdot \left(\frac{1}{5}\right)^x = 30 \cdot 2^x$ using graphs:

a. Graph the equations $y = 3 \cdot \left(\frac{1}{5}\right)^x$ and $y = 30 \cdot 2^x$ in the same coordinate plane.

b. In how many points do these graphs intersect? What do these points of intersection represent?

c. Using a table of values for both equations, find approximately the X-coordinate of the point where the graphs intersect.

d. Using technology, find exactly the coordinates of the point where the graphs of the equations intersect.

e. What is the solution of the equation $3 \cdot \left(\frac{1}{5}\right)^x = 30 \cdot 2^x$? What is the solution of the system of equations: $\begin{cases} y = 3 \cdot \left(\frac{1}{5}\right)^x \\ y = 30 \cdot 2^x \end{cases}$? Explain why the X-coordinates of the points where the graphs of the equations $y = 3 \cdot \left(\frac{1}{5}\right)^x$ and $y = 30 \cdot 2^x$ intersect are the solutions of the equation $3 \cdot \left(\frac{1}{5}\right)^x = 30 \cdot 2^x$.

6. Solve the equation $3^x = -x + 4$ using graphs:

    a. Graph the equations $y = 3^x$ and $y = -x + 4$ in the same coordinate plane.

    b. In how many points do these graphs intersect? What do these points of intersection represent?

    c. Using a table of values for both equations, find approximately the X-coordinate of the point where the graphs intersect.

    d. Using technology, find exactly the coordinates of the point where the graphs of the equations intersect.

    e. What is the solution of the equation $3^x = -x + 4$? What is the solution of the system of equations: $\begin{cases} y = 3^x \\ y = -x + 4 \end{cases}$? Explain why the X-coordinates of the points where the graphs of the equations $y = 3^x$ and $y = -x + 4$ intersect are the solutions of the equation $3^x = -x + 4$.

7. Solve the equation $\frac{3x+1}{2} = \frac{1}{2} \cdot 7^x$ using graphs:

    a. Graph the equations $y = \frac{3x+1}{2}$ and $y = \frac{1}{2} \cdot 7^x$ in the same coordinate plane.

    b. In how many points do these graphs intersect? What do these points of intersection represent?

    c. Using a table of values for both equations, find approximately the X-coordinate of the point where the graphs intersect.

    d. Using technology, find exactly the coordinates of the point where the graphs of the equations intersect.

    e. What is the solution of the equation $\frac{3x+1}{2} = \frac{1}{2} \cdot 7^x$? What is the solution of the system of equations: $\begin{cases} y = \frac{3x+1}{2} \\ y = \frac{1}{2} \cdot 7^x? \end{cases}$ Explain why the X-coordinates of the points where the graphs of the equations $y = \frac{3x+1}{2}$ and $y = \frac{1}{2} \cdot 7^x$ intersect are the solutions of the equation $\frac{3x+1}{2} = \frac{1}{2} \cdot 7^x$.

F-IF.1 *Understand that a function from one set (called the domain) to another set (called the range) assigns to each element of the domain exactly one element of the range. If f is a function and x is an element of its domain, then f(x) denotes the output of f corresponding to the input x. The graph of f is the graph of the equation y = f(x).*

1. If $f: A \rightarrow B$ is a function, and the domain $A = \{1, 2, 3, 4\}$ and $f(x) = x^2 + 2x - 3$, find set B (the range). Complete a table and graph the function f.

2. If $f: A \rightarrow B$ is a function, and the range $B = \{1, 2, 3, 4\}$ and $f(x) = 2x - 5$, find set A (the domain). Complete a table and graph the function f.

3. If $f: \{2, 3, 5\} \rightarrow \{a, b, c\}$, $f(x) = 4x - 7$ is a function, what are the values of a, b, c? How many combinations are there?

4. If $f: A \rightarrow B$ is a function, and the domain $A = \{1, 2, 3\}$ and the range $B = \{1, 2\}$, how many functions can you define from A to B?

5. If $f(x) = x^3 - 4x + 1$, $f: \{1, 2, 3\} \rightarrow B$ is a function, graph the function f.

6. Given the diagrams below:

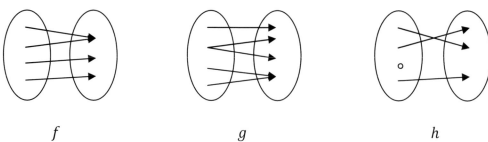

$$f \qquad\qquad\qquad g \qquad\qquad\qquad h$$

Which of the diagrams describe a function? Explain.

7. If $f: \{1, a\} \rightarrow \{1, b\}$ is a function with $f(x) = 3x - 5$, what are the values of a and b? Complete a table and graph the function f.

8. Graph the function $f: A \rightarrow B$, $f(x) = 2x - 1$, in the following situations:
   a. $A = \boldsymbol{R}$ (the set of all real numbers)
   b. $A = \{2, 3, 4\}$
   c. $A = \{x | 2 \leq x < 5\}$
   d. $A = \{x | x \geq 2\}$
   e. $A = \{x | -1 \leq x \leq 2 \text{ or } 4 < x \leq 8\}$

9. Complete the tables and then graph each function:
   a. $f(x) = -3x^2 + 4$

X	-2	-1	0	1	2	3
f(x)						

   b. $f(x) = 4 \cdot |x| - 3$

X	-3	-1	1	3	5
f(x)					

c. $f(x) = 2x - 5$

X	-4		-1		5	7
f(x)		2		-3		

d. $f(x) = -x + 3$

X	-1	3		
f(x)			5	-7

10. What is the value of $a + b + c + d$, if $f:\{-5, a, 2, b\} \to \{c, 4, d, -1\}$, $f(x) = -2x + 9$ is a function?

11. If $f: A \to B$, $f(x) = \begin{cases} 2x + 3, if \ x < 3 \\ -x + 5, if \ x \geq 3 \end{cases}$ is a function and $A = \{-1, 0, 1, 2, 3, 4\}$ is the domain, complete a table and graph the function f.

12. The function $r(x) = 60x$ represents the number of words $r(x)$ you can type in x minutes. How many words can you type in 10 minutes?

13. Sound travels about 120,000 miles per second. The function $d(t) = 120000t$ gives the distance $d(t)$, in miles, that sound travels in t seconds. How far does sound travel in 20 seconds?

14. A truck can travel 14 miles for each gallon of gasoline. The function $d(x) = 14x$ represents the distance $d(x)$, in miles, that the truck can travel with x gallons of gasoline. How many miles can the truck travel with 10 gallons of gasoline?

15. A shop rents bikes for a $12 equipment fee plus $4 per hour, with a maximum cost of $30 per day. Express the number of hours x and the cost y as a function in table form, and find the cost to rent a bike for 3, 4, 5, 6, 7, 8 or 9 hours.

16. The *input* of a function is the *independent* variable and the *output* of a function is the *dependent* variable. Identify the independent and dependent variables in each situation:
    a. During the winter, more electricity is used when the temperature goes down, and less is used when the temperature rises;
    b. The cost of shipping an envelope is based on its weight;
    c. The time it takes Felix to get home depends on the speed he walks.

17. Identify the independent and dependent variables. Write an equation in function notation for each situation:
    a. A lawyer's fee is $150 per hour for her services;
    b. The admission fee to an amusement park is $10. Each ride costs $3;
    c. Marcos buys tomatoes that costs $1.90 per pound;
    d. An ice rink charges $5 for skates and $1.50 per hour.

18. A swimming pool containing 20,000 gallons of water is being drained. Every hour, the volume of the water in the pool decreases by 750 gallons.

   a. Write an equation to describe the volume V of water in the pool after h hours.

   b. How much water is in the pool after 2 hours?

   c. Create a table of values showing the volume of the water in gallons in the swimming pool as a function of the time in hours and graph the function.

19. The temperature of a chemical substance that started at 20 °C is increasing by 2 °C per hour. Write a function that describes the temperature of the substance over time. Graph the function to show the temperature over the first 8 hours.

20. Suppose f is a function.

   a. If $8 = f(-2)$, find the coordinates of a point on the graph of f.

   b. If 7 is a solution of the equation $f(t) = 3$, find a point on the graph of f.

F-IF.2 *Use function notation, evaluate functions for inputs in their domains, and interpret statements that use function notation in terms of a context.*

1. If $f(x) = 2x - 5$ is a function, evaluate $f(4)$.
2. If $f(x) = 3 \cdot 2^x$ is a function, evaluate $f(2)$.
3. If $f(x) = x^2 - 2x + 3$ is a function, evaluate $f(-1)$.
4. If $f(x) = \frac{1}{2} \cdot x - \frac{3}{4}$ is a function, evaluate $f(-4)$.
5. If $f(x) = x^2 - \frac{1}{3} \cdot x + 4$ is a function, evaluate $f(-\frac{1}{2})$.
6. If $f(x) = \frac{4x-1}{x+5}$ is a function, evaluate $f(-4)$.
7. If $f(x) = (2x - 1) \cdot (3x + 5) + 9$ is a function, evaluate $f(-2)$.
8. If $f(x) = 2 \cdot |3x + 4| - 5$ is a function, evaluate $f(-5)$.
9. If $f(x) = \begin{cases} 4x - 3, & \text{if } x \leq 2 \\ -2x + 5, & \text{if } x > 2 \end{cases}$ is a function, evaluate $f(-3), f(4), f(-1), f(2), f(0),$ $f(1), f(3)$.
10. If $f(x) = 2x - 5$ is a function, solve for x:
$$3 \cdot f(x) - 2 = 7$$
11. If $f(x, y) = x^2 - 2xy + 3y^2$ is a function, evaluate $f(-1, 2)$.
12. The height, $h(t)$, in yards of an object thrown into the air with an initial upward velocity of 25 yards per second is given by the function $h(t) = -2t^2 + 25t$, where t is the time in seconds. What is the height of the object after 9 seconds?
13. Suppose a ball is dropped from a height of 7 meters and bounces to 87% of its previous height after each bounce. Using the function $h(n) = 7 \cdot (0.87)^n$, where n represents the number of bounces and $h(n)$ represents the maximum height of the ball after the nth bounce, what is maximum height of the ball after the 12th bounce?
14. The value of Mr. Holliday's car x years after its purchase is given by the function $V(x) = 14,000(0.91)^x$. What was the value of Mr. Holliday's car 3 years after its purchase?
15. The function $p(t) = 2.7 \cdot (1.08)^t$ gives a city's population $p(t)$ (in millions), where t is the number of years since 1998. According to this function, what was the population of the city in 2010?
16. Evaluate $h(x) = \frac{3}{4} \cdot (5 - 8x) + 3x$ when $x = \frac{3}{8}$.
17. The annual tuition at a college since 1980 is modeled by the equation $T = 4000(1.05)^x$, where T is the tuition cost and x is the number of years 1980.
    a. What was the tuition cost in 1980?
    b. What is the annual percentage of tuition increase?
    c. Find the tuition cost in 2010.

18. A swimming pool containing 16,000 gallons of water is being drained. Every hour, the volume of the water in the pool decreases by 800 gallons.

     a. Write an equation to describe the volume V of water in the pool after h hours.

     b. How much water is in the pool after 5 hours?

     c. Create a table of values showing the volume of the water in gallons in the swimming pool as a function of the time in hours and graph the function.

19. The temperature of a chemical substance that started at 10 °C is increasing by 3 °C per hour. Write a function that describes the temperature of the substance over time. Graph the function to show the temperature over the first 6 hours.

F-IF.3 *Recognize that sequences are functions, sometimes defined recursively, whose domain is a subset of the integers.*

1. What is a sequence of numbers?
2. How can you define a sequence?
3. Give an example of a sequence defined recursively. Explain the meaning of the word "recursive".
4. Give an example of a sequence defined explicitly.
5. If $a_n = 2n + 5$ is a sequence, find a_3, a_4, a_{10}.
6. If $x_n = \frac{n^2-2n+3}{n^2+1}$ is the nth term of a sequence, find x_1, x_3, x_4.
7. If $x_n = 3 \cdot x_{n-1} + 5$ is a sequence defined recursively and $x_1 = 9$, find x_2, x_5, x_6.
8. If $x_{n+1} = x_n{}^2 + x_n - 5$ is a sequence defined recursively and $x_3 = 2$, find x_4, x_5, x_7.
9. If $x_{n+1} = 2 \cdot x_n - 3 \cdot x_{n-1}$ is a sequence defined recursively and $x_0 = 4, x_1 = 5$, find x_2, x_3, x_4.
10. If $b_{n+1} = 0.5 \cdot b_n$ is a sequence defined recursively and $b_4 = 9$, find b_7.
11. If $c_{n+1} = c_n + 3$ is a sequence defined recursively and $c_1 = 2$, find $c_2, c_3, c_4, c_{50}, c_{100}$.
12. If $c_{n+1} = 2 \cdot c_n$ is a sequence defined recursively and $c_1 = \frac{1}{2}$, find c_2, c_3, c_4, c_{10}.
13. If $x_{n+1} = \sqrt{x_n + 6}$ is a sequence defined recursively and $x_1 = 3$, find x_2, x_3, x_4, x_{100}.
14. A single virus cell is placed in a test tube and splits in two after one minute. After two minutes, the resulting two virus cells split in two, creating four virus cells. This process continues for one hour until test tube is filled up. How many virus cells are in the test tube after 6 minutes? How about 14 minutes? Write a recursive rule to find the number of virus cells in the test tube after n minutes. Convert this rule into explicit form. How many virus cells are in the test tube after one hour?
15. Write a recursive formula for each sequence:
    a. 19, 15, 11, 7,...
    b. 2, 6, 18, 54, 162,...
    c. $7, -2, -11, -20, \dots$
    d. $5, 19, 33, 47, 51, \dots$
    e. $243, 81, 27, 9, \dots$
    f. $80, -120, 180, -270, \dots$
16. For each recursive formula, write an explicit formula:
    a. $a_1 = 7, a_n = a_{n-1} - 3$
    b. $a_1 = 64, a_n = \frac{1}{4} \cdot a_{n-1}$
    c. $a_1 = 400, a_n = 0.4 \cdot a_{n-1}$

17. For each explicit formula, write a recursive formula:

   a. $a_n = 2 \cdot 3^{n-1}$

   b. $a_n = -4n + 15$

   c. $a_n = 7n + 1$

   d. $a_n = 6 \cdot 5^{n-1}$

18. The number of bacteria in a culture can be represented by the rule $N_t = 3.18 \cdot N_{t-1}$. In the formula, N_t is the number of bacteria at the end of t minutes, and N_{t-1} is the number of bacteria at the end of $t - 1$ minutes. There are 12,106 bacteria in the culture at the end of 9 minutes. How many bacteria will be in the culture at the end of 10 minutes?

19. Felix received an email that he forwarded to three of his friends. Each of his friends forwarded the email to three more friends, and so on.

   a. Find the first five terms of the sequence representing the number of people who receive the email;

   b. Write a recursive formula for the sequence;

   c. If the number of emails that Felix received represents x_1, find x_6.

20. A certain culture of bacteria increases by 20% every five hours. A scientist places 12 grams of the bacteria in a culture dish. Write the explicit and recursive formulas for the sequence formed by the growth of the bacteria.

21. The balance B_{n+1} in Mr. Ackerman's savings account at the end of a year is calculated by the equation $B_{n+1} = 1.028 \cdot B_n$, where B_n is the balance at the end of the previous year. Mr. Ackerman made a deposit to open the account 4 years ago. He has not made any additional deposits or withdrawals since. The balance at the end of 2 years was $9,537.14. What is the balance at the end of 4 years?

22. The number of bacteria in a culture can be represented by the formula $N_t = 3.7 \cdot N_{t-1}$. In the formula, N_t is the number of bacteria at the end of t minutes, and N_{t-1} is the number of bacteria at the end of $t - 1$ minutes. There are 8,417 bacteria in the culture at the end of 5 minutes. How many bacteria will be in the culture at the end of 9 minutes?

F-IF.4 For a function that models a relationship between two quantities, interpret key features of graphs and tables in terms of the quantities, and sketch graphs showing key features given a verbal description of the relationship. Key features include: intercepts; intervals where the function is increasing, decreasing, positive, or negative; relative maximums and minimums; symmetries.

1. Find the X and Y intercepts of the following functions:
   a. $f(x) = 2x - 4$
   b. $f(x) = x^2 + 2x - 3$
   c. $f(x) = 3 \cdot e^{2x}$
   d. $f(x) = -3x + 9$
   e. $f(x) = -5x^2 + 3x + 2$
   f. $f(x) = 2 \cdot e^{-x}$
   g. $f(x) = \dfrac{2x+4}{x-3}$
   h. $f(x) = \sqrt{x^2 + 5x - 6}$
   i. $f(x) = \dfrac{x^2+2x-15}{x^2-1}$

2. Find the X and Y intercepts of the function: $f(x) = \begin{cases} 2x - 6, & \text{if } x \geq 1 \\ 3x + 9, & \text{if } x < 1 \end{cases}$

3. Find the X and Y intercepts of the function: $f(x) = \begin{cases} x - 1, & \text{if } x \geq -2 \\ -2x - 8, & \text{if } x < -2 \end{cases}$

4. Find the X and Y intercepts of the function: $f(x) = \begin{cases} x^2 - 4, & \text{if } x \leq -1 \\ -x^2 + 9, & \text{if } x > -1 \end{cases}$

5. Given a graph:
   a. What are the X intercepts of a function f ?
   b. What is the Y intercept of a function f ?
   c. Where is a function f increasing (decreasing)? How do you interpret it?
   d. What are the relative minimums (maximums) of a function f ? How do you interpret it?
   e. Where is a function f positive? Where is a function f negative? How do you interpret it?
   f. When is a function f symmetric (over a line or a point)?

6. Given the graph below:

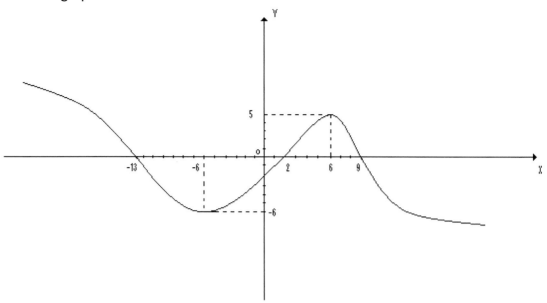

    a. What are the X intercepts of the function *f*? What is the Y intercept?

    b. Can a function have more than one Y intercept? Explain your answer.

    c. Where is the function *f* increasing? Decreasing?

    d. Where is the function *f* positive? Negative?

    e. Specify the relative minimums and maximums of the function *f*.

    f. Is there any symmetry? Explain your answer.

7. Sketch the graph of a function *f* knowing the following features:

    a. X intercepts: $-2,\ 4,\ 12$

    b. Y intercept: 3

    c. *f* is increasing when $-9 < x < -1$ and $x > 8$

    d. *f* is decreasing when $x < -9$ and $-1 < x < 8$

    e. *f* has a maximum of 12 and two relative minimums of -6 and -4

8. Find all the key features (X intercepts, Y intercept, relative minimums, relative maximums, where is the function increasing/decreasing, positive or negative) for the following functions (using maybe the graphing calculator):

    a. $f(x) = 3x + 6$

    b. $f(x) = 2 \cdot e^{4x}$

    c. $f(x) = x^2 + 3x - 4$

    d. $f(x) = -x + 5$

    e. $f(x) = x^2 - 3x + 2$

    f. $f(x) = -4 \cdot e^{3x}$

    g. $f(x) = -x^2 + 4x + 5$

h. $f(x) = \dfrac{2x+3}{x-7}$

i. $f(x) = \dfrac{x^2+2x-3}{x^2-4}$

j. $f(x) = \dfrac{x+1}{x^2-5x+6}$

9. Sketch graphs of functions with the following characteristics:

    a. The graph is linear with an X intercept at -4; the function is positive for $x < -4$ and negative for $x > -4$; the Y intercept is -1

    b. The nonlinear graph has X intercepts at -3 and 1 and Y intercept at -2; the function has a relative minimum of -5 at $x = -2$; the graph is decreasing for $x < -2$ and increasing for $x > -2$.

10. Estimate and interpret the key features of the following graphs (Intercepts, any symmetry, the intervals where the function is positive or negative, the intervals where the function is increasing or decreasing, the coordinates of any relative minimum or maximum, the end behavior of the graph)

    a. Cell Phone Company Service

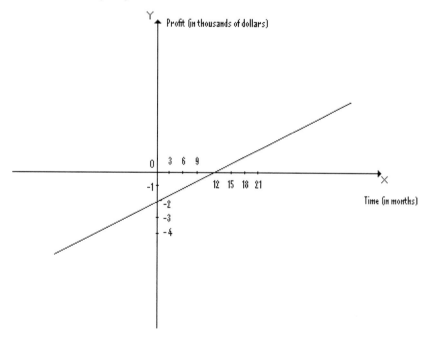

b. Computer Value Depreciation

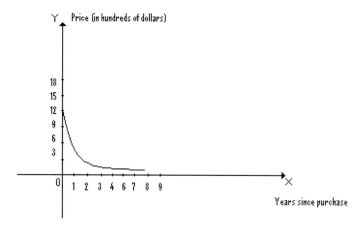

c. Company Profit

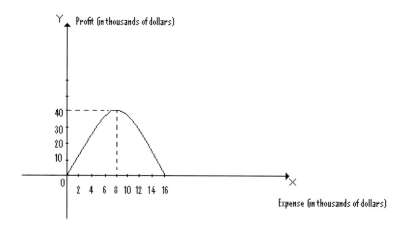

d. Social Network Usage

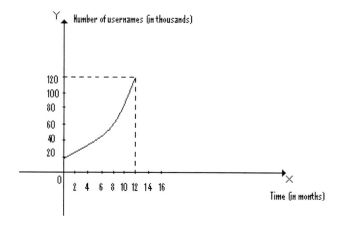

F-IF.5 *Relate the domain of a function to its graph and, where applicable, to the quantity relationship it describes.*

1. Graph each function for the given domain:
   a. $f(x) = -2x + 3$, $D = \{-2,0,1,3\}$
   b. $f(x) = x^2 + 2x - 3$, $D = \{-4,-3,-1,0,1,2\}$
   c. $f(x) = 3 \cdot 2^x$, $D = \{-1,0,1,2\}$

2. For each function, determine whether the given points are on the graph:
   a. $f(x) = \frac{2}{3}x - 5$, $A(-3,-7), B(3,-3), C(6,-4)$
   b. $f(x) = x^2 + 3x$, $A(1,4), B(-2,-3), C(-2,-2)$

3. The gas tank in Felix's truck holds 25 gallons, and the truck can travel 15 miles for each gallon of gas. When Felix begins with a full tank of gas, the function $f(x) = -\frac{1}{15}x + 25$ gives the amount of gas $f(x)$ that will be left in the tank after traveling x miles (if he does not buy more gas).
   a. What is the practical domain and range for this function?
   b. Graph the function on the practical domain.

4. A bank employee notices an abandoned checking account with a balance of $240. If the bank charges a $5 monthly fee for the account, the function $b = 240 - 5m$ shows the balance b in the account after m month.
   a. What is the practical domain and range for this function?
   b. Graph the function on the practical domain.
   c. Find the intercepts. What does each intercept represent?
   d. When will the bank account balance be 0?

5. The function $w = 125c + 20000$ represents the total weight w, in pounds, of a truck that carries c cubic feet of bricks. If the capacity of the truck is about 200 cubic feet, graph the function on the practical domain. What is the practical range of the function? What does the Y intercept represent in the context?

6. The cost C, in dollars, for delivered sandwiches depends on the number S of sandwiches ordered. This situation is represented by the function rule: $C = 5 + 9P$. Graph the function and interpret the practical domain and the Y intercept.

7. An arrow is launched from 2 meters above the ground at time $t = 0$. The function that models this situation is given by $h(t) = -\frac{1}{16}t^2 + \frac{7}{8}t + 2$, where t is the time measured in seconds and h is the height above the ground measured in meters.
   a. What is the practical domain for t in this context?
   b. What is the height of the arrow three seconds after it was launched?
   c. What is the maximum value of the function and what does it mean in context?
   d. When is the arrow 3 meters above the ground?
   e. When is the arrow 1 meter above the ground?

    f. What are the intercepts of this function? What do they mean in the context of this problem?

    g. What are the intervals of increase and decrease on the practical domain? What do they mean in the context of this problem?

8. Felix has a gift card to use at a gas station. The function $d = 50 - 3.25g$ represents the remaining dollars d on the gift card after buying g gallons of gas. Find the zero of this function and describe what this value means in this context.

9. A bus is driving at 45 miles per hour toward a vacation resort that is 360 miles away. The function $d = 360 - 45t$ represents the distance d from the resort to the bus, t hours after it has started driving. Find the zero of this function and describe what this value means in this context.

F-IF.6 Calculate and interpret the average rate of change of a function (presented symbolically or as a table) over a specified interval. Estimate the rate of change from a graph.

1. The table shows the balance of a bank account on different days of the month.

Day	1	5	10	15	22	30
Balance	2500	2000	1730	1300	1000	720

   a. Find the rate of change for each time interval.

   b. During which time interval did the balance decrease at the greatest rate?

2. The table shows the price of gasoline per gallon in different years.

Year	1980	1985	1990	1995	2005	2010
Price per gallon	$1.25	$1.42	$1.68	$1.73	$2.80	$3.40

   a. Find the rate of change in cost for each time interval.

   b. During which time interval did the cost increase at the greatest rate?

3. Graph the data below, find the rates of change and interpret them:

Time(hours)	1	2	3	4	5
Distance(Miles)	6	10.2	13.5	16	18.8

4. The table below shows the approximate number of sharks in the west side of the Atlantic Ocean after the year 1975.

Time(Years)	1975	1985	1995	2005
Number of sharks	245	576	1376	2436

   Graph the data and show the rates of change.

5. The table below shows the average salary in North Carolina in different years:

Year	1970	1980	1990	2000	2010
Average salary	$900	$1300	$1600	$2100	$2400

   a. Find the rate of change for each ten-year time period.

   b. During which time period did the average salary increase the most? Explain what does the rate of change for this time period mean.

6. Use rates of change to determine whether the following functions are linear or non-linear. (*If the function is linear then the rate of change of the function is the same as the slope of the line that represents the graph of the function*)

a.

x	-2	1	4	8	12
y	5	3	-2	10	15

b.

x	-3	1	5	9	12
y	7	15	23	31	37

c.

x	-6	-1	2	7	8
y	-8	7	16	31	34

d.

x	1	2	3	4	5	6
y	10	12	15	19	24	30

7. Look at the graph below and estimate between what points there is the highest rate of change. (*If the line is steeper then the rate of change is greater*)

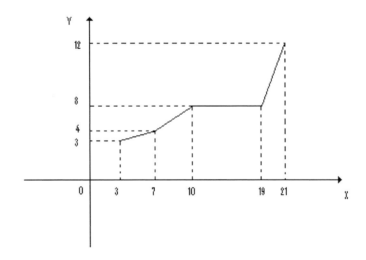

8. For a function $f(x)$, whose graph contains the points (x_1, y_1) and (x_2, y_2) *the average rate of change* over the interval $[x_1, x_2]$ is the slope of the line through (x_1, y_1) and (x_2, y_2). If $f(x) = 2x^2 - 3x + 1$, find the average rate of change over the interval $[0,3]$.

9. What is the average rate of change over the interval $[1,5]$ for the function graphed below:

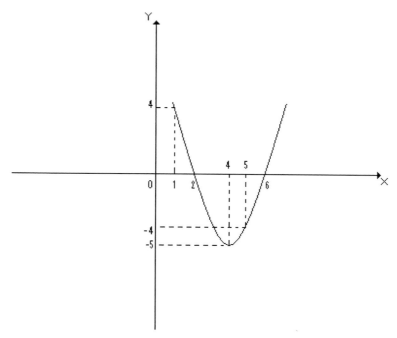

10. Find the average rate of change for the following functions over the specified intervals:
   a. $f(x) = -3x + 5$; interval $[1,7]$
   b. $f(x) = 4x - 9$; interval $[-2,5]$
   c. $f(x) = 3 \cdot 2^x$; interval $[0,2]$
   d. $f(x) = x^2 - 4x + 5$; interval $[-4,9]$
   e. $f(x) = -x^2 + 4$; interval $[-2,1]$
11. Let $f(x) = 3x - 17$ be a function.
   a. Calculate the average rate of change over the interval $[-1,2]$.
   b. Calculate the average rate of change over the interval $[3,10]$.
   c. What did you notice about the answers for part (a) and part (b)? Explain. Is there a constant rate of change for this function?
   d. Do all functions have a constant rate of change? Explain using examples.
   (For linear functions the rate of change is equal with the slope of the line that graphs the function)
12. Felix has $5000 to invest in one of the two financial plans offered by a bank. Plan A offers to increase his principal by $180 each year, while plan B offers to pay 3.8% interest compounded quarterly. The amount of each investment after t years, is given by: $A = 5000 + 180t$ and $B = 5000 \cdot (1.0095)^{4t}$ respectively.
   What plan should Felix choose if he invests his money for 5 years? What if he invests his money for 10 years? Explain your reasoning, using function values, the average rate of change and the graphs of the equations.

F-IF.7 *Graph functions expressed symbolically and show key features of the graph, by hand in simple cases and using technology for more complicated cases.*

> *a. Graph linear and quadratic functions and show intercepts, maxima, and minima.*

> *b .Graph exponential functions, showing intercepts and end behavior.*

1. Graph $2x + 5y = 10$ using the X and Y intercepts.
2. Graph $y = \frac{1}{2}x - 4$ using a table.
3. Felix earns a monthly salary of $2500 and a commission of $250 for each painting he sells.
    a. Graph an equation that represents how much Felix earns in a month in which he sells x paintings.
    b. Use the graph to estimate the number of paintings Felix needs to sell in order to earn $5000.
4. Felix, Emilia and Nancy are going to an amusement park. The function $A = 200 - 15R$ represents the amount of money, A, they have left after R rides.
    a. What are the intercepts of this function and explain what do they represent in this context?
    b. What is the practical domain and range of this function?
5. Graph the following functions and find the X and Y intercepts of the graphs:
    a. $f(x) = -2x + 4$
    b. $f(x) = x - 3$
    c. $y = 3x + 6$
    d. $y = -x + 1$
    e. $f(x) = 5x + 10$
    f. $y = \frac{1}{3}x - 2$
    g. $f(x) = -\frac{2}{5}x + 10$
6. For the following Quadratic Functions find the Vertex, the X intercepts, the Y intercept and graph each function. After graphing, find the axis of symmetry, the minimum or maximum, the domain and the range for each function:
    a. $f(x) = x^2 + 2x - 3$
    b. $f(x) = 2x^2 - 8x + 7$
    c. $f(x) = -x^2 + 2x - 10$
    d. $y = -3x^2 + 12x - 5$
    e. $y = x^2 + 8x + 20$
    f. $f(x) = x^2 - 4$

g. $y = 2x^2 + 6x$

h. $f(x) = 2x^2 + 5$

i. $y = -x^2 + 3x$

7. A football is kicked up from ground level at an initial upward velocity of 20 feet per second. The function $h(t) = -4t^2 + 20t$ gives the height of the football after t seconds.

    a. Graph the function h.

    b. What is the height of the football after 2 seconds?

    c. When is the ball 16 feet high?

    d. What are the zeros (X intercepts) of the function and what do they represent?

    e. What is the maximum height of the football and when is it reached?

    f. How long was the football in the air?

8. Graph the following exponential functions, showing intercepts and end behavior:

    a. $f(x) = 2^x$

    b. $f(x) = 5 \cdot 3^x$

    c. $y = -3 \cdot 4^x$

    d. $f(x) = -\left(\frac{1}{3}\right)^x$

    e. $y = 2 \cdot \left(\frac{1}{6}\right)^x$

    f. $f(x) = 5 \cdot 3^x - 1$

    g. $y = -2 \cdot 5^x + 7$

    h. $f(x) = 3^x - 9$

    i. $y = 3 \cdot 2^x - 6$

    j. $y = 16 \cdot \left(\frac{1}{2}\right)^x$

    k. $f(x) = 4 \cdot 3^{-x}$

    l. $y = \frac{1}{2} \cdot 3^x$

9. A lake has a population of 5000 sardines. Each year the population decreases by 250 sardines. The population in the lake after x years is represented by the function $f(x) = 5000 - 250x$.

    a. Graph the function.

    b. Find the X and Y intercept of the function.

    c. What does each intercept represent?

10. The height in feet of a football that is kicked can be modeled by the function $f(x) = -18x^2 + 72x$, where x is the time in seconds after the ball is kicked.

    a. Graph the function f.

    b. What is the height of the football after 2 seconds?

    c. When is the ball 54 feet high?

    d. What are the zeros (X intercepts) of the function and what do they represent?

    e. What is the maximum height of the football and when is it reached?

    f. How long was the football in the air?

11. Blake has a gift card to use at a gas station. The function $d = 100 - 3.75g$ represents the remaining dollars d on the gift card after buying g gallons of gas. Find the zero of this function and describe what this value means in this context.

12. A bus is driving at 45 miles per hour toward a vacation resort that is 270 miles away. The function $d = 270 - 45t$ represents the distance d from the resort to the bus, t hours after it has started driving. Find the zero of this function and describe what this value means in this context.

F-IF.8 *Write a function defined by an expression in different but equivalent forms to reveal and explain different properties of the function.*

    a. *Use the process of factoring a quadratic function to show zeros and interpret these in terms of a context.*
    b. *Use the properties of exponents to interpret expressions for exponential functions.*

A-SSE.3 *Choose and produce an equivalent form of an expression to reveal and explain properties of the quantity represented by the expression.*

    a. *Factor a quadratic expression to reveal the zeros of the function it defines.*

1. A college's tuition has been increasing 3% each year since 1980. If the tuition cost in 1980 was $7,000, write a function for the amount of the tuition x years after 1980. Find the cost of tuition for this college in 2014.

> **The equation for compound interest is $A = P \cdot \left(1 + \frac{r}{n}\right)^{n \cdot t}$, where P is the principal (initial amount), r is the annual interest rate (expressed as a decimal), n is the number of times the interest is compounded each year, t is the time in years, A is the amount after t years.**

2. Felix deposited $3,000 in 2009 into a savings amount at a bank that offers 6% per year compounded monthly. How much money will Felix have in 2014?
3. Emilia deposited $7,800 in 2006 into a savings account at a bank that offers 4% per year compounded quarterly. How much money will Emilia have in 2015?
4. Nancy deposited $4,300 in 2010 into a savings amount at a bank that offers 5.3% per year compounded semiannually. How much money will Nancy have in 2018?
5. Donald bought a new car for $28,999. The price of the car depreciated each year with 14%.
   a. Write a function that models the price of the car since it was bought.
   b. What is the price of the car after 10 years?
6. Gabrielle is a doctor and her starting salary in 2010 was $254,000 per year. According to her contract, she receives a 1.7% increase in her salary every year. How much will Gabrielle earn in 2015?
7. The number of students enrolled at a College in 1925 was 500 and it has been growing 4% each year.
   a. Write a function rule that describes the number of students enrolled at this college, x years after 1925.
   b. What is the number of students enrolled in 2012?
8. In the years from 2005 to 2015, the population of New York is expected to model according to the function: $f(x) = 4000000 \cdot (0.85)^x$, where x represents the number of years since 2005. What is the annual percentage rate of change of the population?

9. What are the zeros of the following functions:
    a. $f(x) = x^2 + 2x - 3$
    b. $f(x) = 6x^2 + 5x - 4$
    c. $f(x) = 3x^2 + x - 10$
10. Find the Quadratic function that has the zeros 1 and -4.
11. The annual tuition at a private school since 2004 is modeled by the function $f(x) = 800 \cdot (1.09)^x$, where x is the number of years since 2004.
    a. What was the tuition cost in 2004?
    b. What is the annual percentage of tuition increase?

12. The number of bacteria in a culture changes every year according to the function $f(x) = 100 \cdot (1.7)^x$, where x is the number of years since the bacteria appeared.
    a. What was the initial number of bacteria in the culture?
    b. What is the annual percent of change in the number of the bacteria?

13. Frank deposited a certain amount of money into a bank. After x years, Frank's amount can be calculated by the function $f(x) = 4000 \cdot (1.005)^{12x}$.
    a. What was the amount of money that Frank invested in the beginning?
    b. Interpret the function and describe the financial plan that the bank offered to Frank.

14. A company earns a weekly profit of P dollars by selling x items according to the function: $P(x) = -x^2 + 25x - 100$
    a. Find the zeros of the function P and interpret these zeros in the context of the problem.
    b. How many items does the company have to sell each week to maximize the profit?

15. Bob deposited a certain amount of money into a bank. After x years, Bob's amount can be calculated by the function $f(x) = 3000 \cdot (1.02)^{4x}$.
    a. What was the amount of money that Bob invested in the beginning?
    b. Interpret the function and describe the financial plan that the bank offered to Bob.

16. Suppose $h(t) = -2t^2 + 4t + 3$ is a function giving the height of a diver above the water (in meters), t seconds after the diver leaves the springboard.
    a. How high above the water is the springboard? Explain your reasoning.
    b. When does the diver hit the water?
    c. At what time on the diver's descent toward the water is the diver again at the same height as the springboard?
    d. When does the diver reach the peak of the dive?

17. A company earns a weekly profit of P dollars by selling x items according to the function: $P(x) = -x^2 + 20x - 96$
    a. Find the zeros of the function P and interpret these zeros in the context of the problem.

b. How many items does the company have to sell each week to maximize the profit?

18. Suppose $h(t) = -3t^2 + 5t + 4$ is a function giving the height of a diver above the water (in meters), t seconds after the diver leaves the springboard.

   a. How high above the water is the springboard? Explain your reasoning.

   b. When does the diver hit the water?

   c. At what time on the diver's descent toward the water is the diver again at the same height as the springboard?

   d. When does the diver reach the peak of the dive?

A common application of exponential decay is half-life. The half-life of a substance is the time it takes for one-half of the substance to decay into another substance.

$$A = P \cdot (0.5)^t$$

P represents the original amount, A represents the final amount and t represents the number of half-lives in a given time period.

19. A certain substance has a half-life of 20 seconds. Find the amount of the substance left from an 80 gram sample after 2 minutes.

20. Ribozyme has a half-life of 5 days. Find the amount of Ribozyme left from a 150 gram sample after 10 weeks.

F-IF.9 Compare properties of two functions each represented in a different way (algebraically, graphically, numerically in tables, or by verbal descriptions).

1. Two phone companies offer the following plans: Company A charges $40 a month and 6 cents per minute of usage, while company B charges according to the linear function represented by the table:

m (minutes)	0	100	200	300	400	500	600
C (cost)	$35	$43	$51	$59	$67	$75	$83

   a. Describe and compare the two plans given above.
   b. If John enjoys talking on the phone, which plan should he choose?

2. Two families are going on a trip. The first family (the Johnson's) are averaging 50 miles per hour, so the function attached to this situation is $D = 50 \cdot t$, where D is the distance covered in t hours. The other family (the Martin's) travel according to the chart:

t (hours)	0	5	8	12
m (miles)	0	200	320	480

   a. Describe and compare the functions that describe the two trips.
   b. Which family travel more miles in the first day of their vacation?

3. Nova Taxi cab company charges a $3 boarding rate in addition to its meter which is $2 for every mile. PRITAX cab company charges according to the function $P = 1.5x + 6$, where x is the number of miles traveled.

   a. Compare the charging policies of the two cab companies.
   b. Which cab company has a better price? Explain your answer.

4. A factory wants to build a wind turbine. A company comes with two projects. Project A costs $20,000 to build the wind turbine and another $150 per day to operate it. The cost for project B is shown in the graph below:

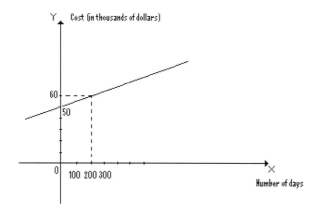

   a. Compare the functions that describe the two projects.
   b. Which project should the factory choose? Explain your answer.

5. The cost of renting an electricity generator A can be represented by the function: $f(x) = 100 + 4.8x$, where x is the number of hours the generator is rented. The cost of renting an electricity generator B is represented in the graph below:

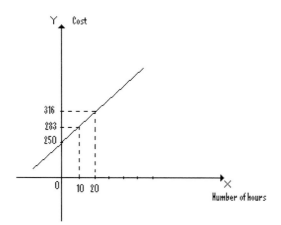

   a. Compare the two functions given above.
   b. Which generator has a better renting price?

6. The height, h, in feet of a football above the ground is given by the function $h(t) = -4t^2 + 32t + 3$, where t is the time in seconds. The height of a soccer ball above the ground is represented in the graph below:

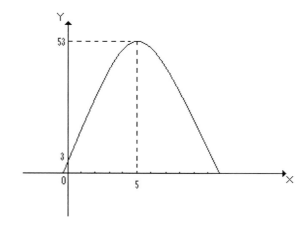

   a. Compare the properties of the two functions given above.
   b. What ball reaches a higher height and which ball stays more in the air?

7. Felix's company weekly revenue in dollars is given by the function $R(x) = 2000x - 2x^2$ where x is the number of items produced during a week. Emilia's company weekly revenue is represented in the graph below:

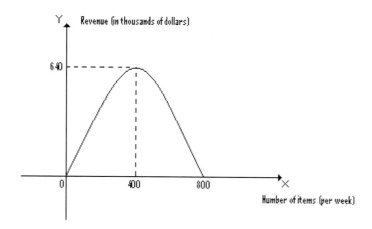

a. What amount of items will produce the maximum revenue in Felix's company? What amount of items will produce the maximum revenue in Emilia's company?

b. Compare the evolution of the revenue of the two companies represented by the functions above.

8. Decide which linear function is increasing at a higher rate:

Function 1 has a X intercept 3 and Y intercept -4,

Function 2 has a graph that includes the points from the table below:

X	1	3	5	7
Y	2	8	14	20

9. Juan is studying population changes in two rural communities. Compare the populations by finding and interpreting the average rates of change over the interval $[0,18]$.

Population 1:

Time (months)	0	6	12	18
Population (thousands)	12	12.4	12.7	13.2

Population 2:

$y = 5.2 \cdot (1.04)^x$, where x is the time in months.

10. The annual tuition at a college A since 2005 is modeled by the function $T(x) = 3000 \cdot (1.03)^x$, where x is the number of years since 2005. The annual tuition at another college B is represented by a graph below:

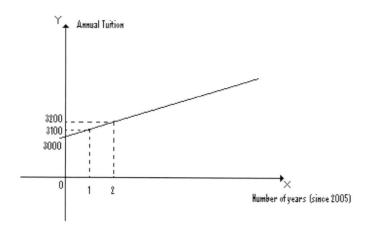

Compare the properties of the two functions and describe the evolution of the tuition cost for the two colleges.

11. The depth of Lake Jake can be described by the function $y = 247 \cdot (0.976)^x$, where x represents the number of years since 2000. The depth of Lake Scott was 150 feet in the year 2000 and the annual increase of the depth is 4%.

   a. What was the initial depth of Lake Jake?

   b. Compare the graphs of the functions that describe the evolution of the depths of the two lakes.

   c. When will the lakes have the same depth?

12. A farm has approximately 500 cows. The owners plan to increase the number of cows. Plan A calls for an increase of 50 cows per year. Plan B calls for a 7% increase each year. Compare the properties of the functions that describe the two plans.

13. Felix wants to invest $5,000 into a financial plan. Three banks offer the following choices as financial investments:

Bank A: $y = 5000 + 200 \cdot x$

Bank B: $y = 5000 + 50x^2 - 200x$

Bank C: $y = 5000 \cdot (1.02)^x$, where y represents the amount of money after x years .

   a. Which plan is better for Felix if he invests his money for 1 year? How about 5 years? How about 10 years?

   b. Compare the three financial plans using tables or graphs and find which financial plan is the best for different time periods.

14. In 1970, three cities, Redville, Pinkville and Blueville had approximately the same population, 3000 people. Since 1970, the three cities developed economically, socially and politically. The population of each city increased in a certain way.

Redville's population increased according to the graph below:

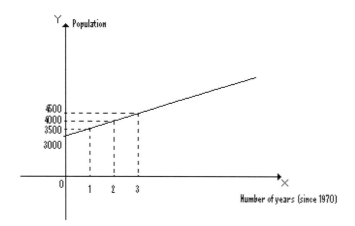

Pinkville's population increased according to the function rule:

$f(x) = -x^2 + 400x + 3000$, where x is the number of years since 1970.

Blueville's population increased 12% each year.

a. Analyze and compare the properties of each function that describes the evolution of the population of each city.

b. What city had the highest increase from 1970 to 1975?

c. What city has more people today?

d. What city is going to have the most people in 2020?

F-BF.1 *Write a function that describes a relationship between two quantities.*

a. *Determine an explicit expression, a recursive process, or steps for calculation from a context.*

b. *Combine standard function types using arithmetic operations.*

1. At the beginning of January, Emilia had some money in her checking account. At the end of each month she deposits enough to double the amount currently in the account. Also, she has to pay off a loan, requiring her to withdraw $20 from the account monthly (at the end of the month immediately after her deposit)

    a. Assuming January is the 1st month , write an equation that describes the amount of money that Emilia has in her account at the end of the nth month, $A(n)$, in terms of the amount of money in Emilia's account at the end of the $(n-1)^{th}$ month, $A(n-1)$.

    b. At the end of April, Emilia had $1,000 left in the account. How much did she have at the beginning of March? How about in the beginning of January?

2. Felix knows that money in an account where interest is compounded semi-annually will earn interest faster than money in an account where interest is compounded annually. He wonders how much interest can be earned by compounding it more and more often. Let's suppose Felix invests $1 at a 100% interest rate.

    a. What will the year-end balance be if the interest is compounded annually?

    b. What will the year-end balance be if the interest is compounded semi-annually?

    c. What will the year-end balance be if the interest is compounded quarterly?

    d. What will the year-end balance be if the interest is compounded monthly?

    Explain how you calculate the year-end balance in each situation.

3. Using the graph below, sketch a graph of the function $s(x) = f(x) + g(x)$ on interval $[-1,8]$

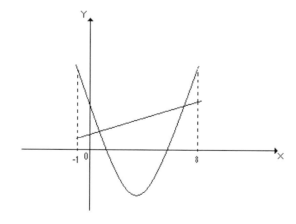

4. The average price of a movie ticket in the year 2000 was $5.39. The average price of a movie ticket in the year 2008 was $9.55.
    a. Assuming the increase is linear, write a function that describes the relationship between the price of the ticket and the number of years since 2000.
    b. What would be the approximate price of a movie ticket in the year 2012?
5. The price of the gas in 1980 was $1.50 per gallon and in 2006 the price of the gas was $2.28 per gallon.
    a. Assuming that the increase is linear, write a function that describes the relationship between the price per gallon and the number of years since 1980.
    b. What would be the price of the gas in 2013?
6. A car that is 5 years old can get to a certain speed in 8 minutes and at 9 years old can get to the same speed in 14 minutes.
    a. Assuming that the decrease is linear, write a function that describes the relationship between the age of the car and the time necessary to get to the same speed.
    b. How fast can it get to the same speed when it is 12 years old?
7. In 2003 Marcos planted a tree that was 3 feet tall. In 2009, the tree was 15 feet tall.
    a. Assuming the growth of the tree is linear, write a function that describes the relationship between the height of the tree and the number of years since 2003.
    b. What is the height of the tree in 2012?
8. At the surface of the ocean, pressure is 1 atmosphere. At 66 ft below sea level, the pressure is 3 atmospheres.
    a. Assuming that the relationship between the depth and the pressure is linear, write a function that describes this relationship.
    b. Predict the pressure at 100 ft below the sea level.
9. Worldwide carbon monoxide emissions are decreasing about 2.6 million metric tons each year. In 2001, carbon monoxide emissions were 79 million metric tons.
    a. Assuming that there is a linear decrease, write a function that describes the relationship between the emissions and the number of years since 2001.
    b. What would be the emissions in 2014?
    c. What were the emissions in 1990?
10. A car rental company charges $19.95 per day plus $0.15 per mile.
    a. Write a function that gives the cost in dollars to travel x miles over a 7-day period.
    b. Calculate the cost in dollars to travel 250 miles over a 2-day period.

11. Felix opens an account with a deposit of $4,000, and he sets up automatic deposits of $200 to the account every month (from another account).

    a. Write a function $A(m)$ to express the amount of money in the account m months after the initial deposit.

    b. Felix plans on spending $800 the first month and $400 in each of the following months for rent and other expenses. Write a function $W(m)$ to express the amount of money taken out of the account each month.

    c. Find $B(m) = A(m) - W(m)$. What does this new function represent?

    d. Will Felix run out of money? If so, when?

12. The value of a work of art is increasing at a rate of 7% per year and its value in 1970 was $50,000.

    a. Write a function that models this situation.

    b. Find the value of the work of art in 2010.

13. The original value of a car is $34,000 and the value decreases by 6% each year.

    a. Write a function that describes the value of the car over the years.

    b. Find the value of the car in 20 years.

14. The population of a small town is increasing at a rate of 2% per year. In 1900 there were 600 people.

    a. Write a function that models this situation.

    b. What was the population of the town in 1950?

    c. What will the population be in 2020?

15. A bacteria cell is placed in a test tube and splits in two after three minutes. After six minutes, the resulting two bacteria cells split in two, creating four virus cells. This process continues for one hour until test tube is filled up. How many bacteria cells are in the test tube after 30 minutes? How about 2 hours? Write a recursive rule to find the number of bacteria cells in the test tube after n minutes. Convert this rule into explicit form. How many bacteria cells are in the test tube after one day?

16. The first term of a sequence is 3, the second term of the sequence is 7, the third term is 11 and so forth.

    a. Determine a recursive process of the sequence.

    b. Write a function that describes the function.

    c. How does the recursive formula relate to the function that describes the sequence?

17. The first term of a sequence is 2, the second term of the sequence is 6, the third term is 18 and so forth.

    a. Determine a recursive process of the sequence.

    b. Write a function that describes the function.

    c. How does the recursive formula relate to the function that describes the sequence?

18. If $f(x) = 2x - 2$ and $g(x) = x^2 + 2x - 3$ are two functions:

    a. Graph the functions f and g.

    b. Find the expression of the function $h(x) = f(x) + g(x)$.

    c. Sketch the graph of the function h using the graphs of the functions f and g.

    d. Graph the function h, using the explicit form found at point (b) and then compare this graph with the graph from point (c).

19. If $f(x) = x^2 + 4x - 5$ and $g(x) = 8$ are two functions:

    a. Graph the functions f and g.

    b. Find the expression of the function $h(x) = f(x) + g(x)$.

    c. Sketch the graph of the function h using the graphs of the functions f and g.

    d. Graph the function h, using the explicit form found at point (b) and then compare this graph with the graph from point (c).

20. If $f(x) = -3x + 9$ and $g(x) = 5$ are two functions:

    a. Graph the functions f and g.

    b. Find the expression of the function $h(x) = f(x) - g(x)$.

    c. Sketch the graph of the function h using the graphs of the functions f and g.

    d. Graph the function h, using the explicit form found at point (b) and then compare this graph with the graph from point (c).

21. If $f(x) = -5$ and $g(x) = 3 \cdot 2^x$ are two functions:

    a. Graph the functions f and g.

    b. Find the expression of the function $h(x) = f(x) + g(x)$.

    c. Sketch the graph of the function h using the graphs of the functions f and g.

    d. Graph the function h, using the explicit form found at point (b) and then compare this graph with the graph from point (c).

22. Felix and Emilia are each working during the summer to earn money in addition to their weekly allowance. Felix earns $20 per hour at his job and his allowance is $40 per week. Emilia earns $15 per hour and her allowance is $40 per week.

  a. Felix wonders who will have more money in a week if they work the same number of hours. Emilia says: "It depends." Explain what she means.

  b. Is there a number of hours worked for which they will have the same income? If so, find that number of hours. If not, why not?

  c. What would happen to the answer to part (b) if Felix gets a raise in his hourly rate? Explain your reasoning.

  d. What would happen to the answer to part (b) if Emilia no longer gets an allowance? Explain your reasoning.

23. In January, a fast growing species of virus is accidentally introduced through a vaccine into a cat's body. It starts to grow and cover the surface of the lungs in such a way that the area covered because of the virus doubles every month. If it continues to grow, the lungs will be totally covered and the cat will suffocate. At the rate it is growing, this will happen in December.

  a. In what month of the year will the lungs be covered half-way?

  b. In June, a veterinarian who runs some tests on the cat warns that the cat will die if the virus is not eliminated from the cat's body. His assistant friend makes fun of him. Why is his friend skeptical of the warning?

  c. In November, because of the cat's condition a very strong drug is applied and the virus is destroyed. However 1% of the surface of the lungs is still covered because of the virus. How well does this solve the problem of the virus in the cat's body?

  d. Write an equation that represents the percentage of the surface area of the lungs that is covered because of the virus as a function of time (in months) that passes since the virus was introduced into the cat's body through the vaccine.

F-BF.2 Write arithmetic and geometric sequences both recursively and with an explicit formula, use them to model situations, and translate between the two forms.

1. Define an arithmetic sequence. Give an example of an arithmetic sequence.
2. Define a geometric sequence. Give an example of a geometric sequence.
3. Find a real-life situation where an arithmetic sequence is involved.
4. Find a real-life situation where a geometric sequence is involved.
5. A bacteria cell is placed in a test tube and splits in two after one minute. After two minutes, the resulting two bacteria cells split in two, creating four virus cells. This process continues for one hour until test tube is filled up.
    a. How many bacteria cells are in the test tube after 6 minutes? How about 14 minutes?
    b. Write a recursive rule to find the number of bacteria cells in the test tube after n minutes. Convert this rule into explicit form.
    c. How many bacteria cells are in the test tube after half-hour?
6. Write a recursive formula for each sequence (identifying first if it's an arithmetic or geometric sequence)
    a. 18, 15, 12, 9,...
    b. 1, 3, 9, 27, 81,...
    c. $7, -3, -13, -23, ...$
    d. $5, 26, 47, 68, 89, ...$
    e. $243, 81, 27, 9, ...$
    f. $80, -120, 180, -270, ...$
7. For each recursive formula, write an explicit formula:
    a. $a_1 = 19, a_n = a_{n-1} - 4$
    b. $a_1 = 1024, a_n = \frac{1}{2} \cdot a_{n-1}$
    c. $a_1 = 200, a_n = 0.2 \cdot a_{n-1}$
8. For each explicit formula, write a recursive formula:
    a. $a_n = 5 \cdot 2^{n-1}$
    b. $a_n = -3n + 14$
    c. $a_n = 5n + 9$
    d. $a_n = 7 \cdot 3^{n-1}$
9. Felix received an email that he forwarded to three of his friends. Each of his friends forwarded the email to three more friends, and so on.
    a. Find the first five terms of the sequence representing the number of people who receive the email.
    b. Write a recursive formula for the sequence.
    c. If the number of emails that Felix received represents x_1, find x_6.

10. A certain culture of bacteria increases by 10% every five hours. A scientist places 50 grams of the bacteria in a culture dish. Write the explicit and recursive formulas for the sequence formed by the growth of the bacteria.

11. At the beginning of January, Emilia had some money in her checking account. At the end of each month she deposits enough to double the amount currently in the account. Also, she has to pay off a loan, requiring her to withdraw \$35 from the account monthly (at the end of the month immediately after her deposit)

    a. Assuming January is the 1st month , write an equation that describes the amount of money that Emilia has in her account at the end of the nth month, $A(n)$, in terms of the amount of money in Emilia's account at the end of the $(n-1)^{th}$ month, $A(n-1)$.

    b. At the end of May, Emilia had \$1,200 left in the account. How much did she have at the beginning of April? How about in the beginning of January?

12. For each recursive formula, write an explicit formula:

    a. $a_1 = -5, a_n = a_{n-1} + 2$

    b. $a_1 = 1, a_n = \frac{1}{10} \cdot a_{n-1}$

    c. $a_1 = 100, a_n = 2 \cdot a_{n-1}$

13. For each explicit formula, write a recursive formula:

    a. $a_n = 100 \cdot \left(\frac{1}{2}\right)^{n-1}$

    b. $a_n = 7n - 2$

    c. $a_n = -2n + 5$

    d. $a_n = 4 \cdot 3^{n-1}$

F-BF.3 *Identify the effect on the graph of replacing f(x) by f(x) + k and f(x + k) for specific values of k (both positive and negative); find the value of k given the graphs. Experiment with cases and illustrate an explanation of the effects on the graph using technology.*

Note: *At this level, limit to vertical and horizontal translations of linear and exponential functions. Even and odd functions are not addressed.*

1. Given the graph of $f(x)$ where does the graph of $f(x + 2)$ translate?
2. Given the graph of $f(x)$ where does the graph of $f(x - 3)$ translate?
3. Given the graph of $f(x)$ where does the graph of $f(x) + 3$ translate?
4. Given the graph of $f(x)$ where does the graph of $f(x) - 5$ translate?
5. Given the graph of $f(x - 3)$ where does the graph of $f(x + 2)$ translate?
6. Given the graph of $f(x)$ where does the graph of $f(x - 3) + 5$ translate?
7. Compare and contrast the graph of each pair of equations:

   a. $f(x) = x$ and $g(x) = x + 2$
   b. $f(x) = 3^x$ and $g(x) = 3^{x+4}$
   c. $f(x) = 2^x$ and $g(x) = 2^x + 5$
   d. $f(x) = 3\left(\frac{1}{2}\right)^x$ and $g(x) = 3\left(\frac{1}{2}\right)^x - 4$
   e. $f(x) = 5 \cdot 2^x$ and $g(x) = 5 \cdot 2^{x-5}$
   f. $f(x) = x + 5$ and $g(x) = x - 3$
   g. $f(x) = \left(\frac{1}{4}\right)^x$ and $g(x) = \left(\frac{1}{4}\right)^x - 2$
   h. $f(x) = 4\left(\frac{2}{3}\right)^x$ and $g(x) = 4\left(\frac{2}{3}\right)^x + 1$
   i. $f(x) = 5 \cdot 2^x$ and $g(x) = 5 \cdot 2^{x-4}$
   j. $f(x) = 10\left(\frac{1}{2}\right)^x$ and $g(x) = 10\left(\frac{1}{2}\right)^{x+3}$
   k. $f(x) = 4^x$ and $g(x) = 4^{x-1} + 5$
   l. $f(x) = 9\left(\frac{1}{3}\right)^x$ and $g(x) = 9\left(\frac{1}{3}\right)^{x+2} - 4$

8. Graph the function $f(x) = 5^x$.
   Then graph the functions $g(x) = 5^{x+2}$ and $h(x) = 5^x + 2$.
   Describe the transformations from the graph of f to the graph of g and h.

9. Graph $f(x)$ and $g(x)$. Then describe the transformation from the graph of f to the graph of g.

   a. $f(x) = x - 3$ and $g(x) = x + 2$ (two possibilities)
   b. $f(x) = 2^x$ and $g(x) = 2^{x-1} + 5$
   c. $f(x) = 8 \cdot \left(\frac{1}{4}\right)^x$ and $g(x) = 8 \cdot \left(\frac{1}{4}\right)^{x+3} - 7$
   d. $f(x) = 4 \cdot 3^x$ and $g(x) = 4 \cdot 3^{x-5} + 2$
   e. $f(x) = x + 4$ and $g(x) = x - 3$ (two possibilities)

10. In which direction is the graph of $f(x) = 4 \cdot e^{x+b}$ translated when b increases?

11. In which direction does the graph of $y = e^x + c$ shift as c decreases?

12. The graph of $f(x)$ is translated to produce the graph of $g(x) = f(x-3) + 4$. In which direction was the graph of f translated?

13. Which equation represents the graph of $f(x) = e^{x+4}$ translated 1 unit to the right and 2 units down?

14. Which equation represents the graph of $f(x) = 2^x$ translated 3 units down?

15. Which equation represents the graph of $f(x) = 5^x$ translated 7 units up?

16. Which equation represents the graph of $f(x) = \left(\frac{1}{4}\right)^x$ translated 5 units to the left?

17. Which equation represents the graph of $f(x) = 3^x$ translated 4 units to the right?

18. Which equation represents the graph of $f(x) = x$ translated 9 units to the left?

19. Which equation represents the graph of $f(x) = 2^{x+1}$ translated 4 units to the left and 3 units up?

20. Which equation represents the graph of $f(x) = 4^x$ translated 7 units down and 5 units to the left?

F-LE.1 *Distinguish between situations that can be modeled with linear functions and with exponential functions*

a. Prove that linear functions grow by equal differences over equal intervals, and that exponential functions grow by equal factors over equal intervals.
b. Recognize situations in which one quantity changes at a constant rate per unit interval relative to another.
c. Recognize situations in which a quantity grows or decays by a constant percent rate per unit interval relative to another.

1. Frank's Bank offers a 4% interest rate per year, while Larry's Bank offers a 3.2% interest per year, compounded monthly. Felix would like to deposit $50,000.
    a. Which bank should Felix choose if he deposits his money for 3 years? What about 10 years?
    b. Write a function rule that gives the balance after x years for Frank's Bank as well as for Larry's Bank.
    c. Compare the two functions and find which bank has a better deal and for what period of time.

2. Emilia has $3,000 in her savings account and she deposits $250 every month.
    a. Determine whether this situation represents a linear model or an exponential model. Explain your reasoning.
    b. Find the function rule that models this situation.
    c. Using the function rule, find how much money will Emilia have in her savings account after 2 years.

3. The price of a new car depreciates 12% each year after it is purchased.
    a. Determine whether this situation represents a linear model or an exponential model. Explain your reasoning.
    b. Find the function rule that models this situation.
    c. Using the function rule, find how much does a car cost after 5 years if the car was bought as new for $25,000.

4. What kind of model does an arithmetic sequence represent? Explain your reasoning using an example.

5. What kind of model does a geometric sequence represent? Explain your reasoning using an example.

6. A phone company charges $20 each month for the service they provide and 4 cents per minute of conversation.
    a. Determine whether this situation represents a linear model or an exponential model. Explain your reasoning.
    b. Find the function rule that models this situation.
    c. Using the function rule, find how much would a person pay per month if he talked 600 minutes.

7. Johnny has to repair his car. The shop charges $25 as an initial fee and $15 per hour of labor.
    a. Determine whether this situation represents a linear model or an exponential model. Explain your reasoning.
    b. Find the function rule that models this situation.
    c. Using the function rule, find how much did Johnny pay if his car was fixed in 6 hours.

8. A swimming pool containing 30,000 gallons of water is being drained. Every hour, the volume of the water in the pool decreases by 250 gallons.
    a. Determine whether this situation represents a linear model or an exponential model. Explain your reasoning.
    b. Find the function rule that models this situation.
    c. Using the function rule, find how much water is in the pool after 10 hours and how long it takes for the swimming pool to be empty.

9. The temperature of a chemical substance that started at 5°C is increasing by 20% every hour.
    a. Determine whether this situation represents a linear model or an exponential model. Explain your reasoning.
    b. Find the function rule that models this situation.
    c. Using the function rule, find the temperature of the substance after 10 hours.
    d. After how many hours does the temperature of the substance exceed 100°C ?

10. The population of a city is increasing 8% every five years.
    a. Determine whether this situation represents a linear model or an exponential model. Explain your reasoning.
    b. Find the function rule that models this situation.
    c. What will the population of the city be in 2015 if the population in 1950 was 20,000 people?

11. The annual tuition at a college since 1920 is increasing 4% every 5 years.

   a. Determine whether this situation represents a linear model or an exponential model. Explain your reasoning.

   b. Find the function rule that models this situation.

   c. Using the function rule, find what was the annual tuition in 2010 if the initial tuition in 1920 was $2,000.

12. Felix has $10,000 to invest in one of the two financial plans offered by a bank. Plan A offers to increase his principal by $500 each year, while plan B offers to pay 2.5% interest compounded annually.

   a. What plan should Felix choose if he invests his money for 5 years? What if he invests his money for 10 years?

   b. Write a function rule that gives the balance after x years for plan A as well as for plan B.

   c. Compare the two functions, determine whether they represent a linear or an exponential model and find which plan is a better deal for Felix and for what period of time.

13. In 1850, two cities, Yellowbourg and Greenbourg had approximately the same population, 200 people. Since 1850, the two cities developed economically, socially and politically. The population of each city increased in a certain way. Yellowbourg's population increased by 30 people every year, while Greenbourg's population increased 6% each year.

   a. Write a function rule that gives the population of each city x years after 1850.

   b. Compare the two functions and determine whether they represent a linear or an exponential model.

   c. What was the approximate population of each city in 1900? How about in 1950?

14. The price of a house increases 2% each year after it is built.

   a. Determine whether this situation represents a linear model or an exponential model. Explain your reasoning.

   b. Find the function rule that models this situation.

   c. Using the function rule, find how much does a house cost after 5 years if the initial price of the house was $120,000.

F-LE.2 *Construct linear and exponential functions, including arithmetic and geometric sequences, given a graph, a description of a relationship, or two input-output pairs (include reading these from a table).*

1. Construct an exponential function knowing that the graph of this function contains the points $A(3, 4)$ and $B(5, 16)$.
2. Construct a linear function knowing that the graph of this function contains the points $A(-2, 5)$ and $B(-7, -1)$.
3. Construct an exponential function given the table below:

x	1	2	3	4
$f(x)$	$\dfrac{3}{2}$	$\dfrac{3}{4}$	$\dfrac{3}{8}$	$\dfrac{3}{16}$

4. Alex is a car dealer. In addition to his salary, he receives a bonus for each car he sells.

Number of cars	5	10	15	20
Bonus	$2,000	$4,000	$6,000	$8,000

   a. Graph this data and determine whether this situation represents a linear model or an exponential model.
   b. Construct a function that models this data.
5. The population of Rabbitville increased exponentially since 1970. In 1970 the population was 3.5 million people and after only 1 year the population increased to 4.375 million people.
   a. Construct the exponential function that represents the evolution of the population in Rabbitville.
   b. According to the exponential model, what is the yearly increase of the population?
6. Construct a linear function that has the graph below:

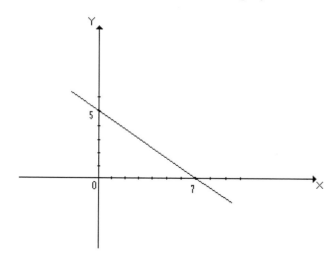

7. Construct an exponential function that has the graph below:

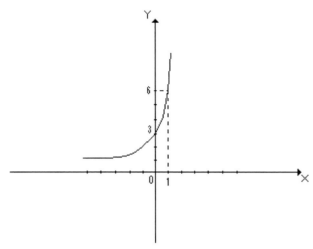

8. The price of a new car depreciates 8% each year after it is purchased. Construct the function that models this situation. Using this function, find the price of a car after 5 years it was purchased if the initial price of the car was $19,500.

9. Tatiana has $5,000 in her savings account and she deposits $300 every month. Construct the function that models this situation. Using this function, find how much money will Tatiana have in her savings account after 3 years.

10. Construct an exponential function knowing that the graph of this function contains the points $P(1, 120)$ and $Q(2, 40)$.

11. Construct a linear function knowing that the graph of this function contains the points $M(3, -7)$ and $N(-9, 2)$.

12. The balance B_{n+1} in Mr. Ackerman's savings account at the end of a year is calculated by the equation $B_{n+1} = 1.028 \cdot B_n$, where B_n is the balance at the end of the previous year. Mr. Ackerman made a deposit to open the account 4 years ago. He has not made any additional deposits or withdrawals since. The balance at the end of 2 years was $9,537.14.

    a. Construct a function that describes the balance in Mr. Ackerman's savings account over the years.

    b. What is the balance at the end of 4 years?

    c. What was the initial amount in Mr. Ackerman's savings account?

13. The number of bacteria in a culture can be represented by the formula $N_t = N_{t-1} + 37$. In the formula, N_t is the number of bacteria at the end of t minutes, and N_{t-1} is the number of bacteria at the end of $t - 1$ minutes. There are 948 bacteria in the culture at the end of 5 minutes.

    a. Construct a function that describes the number of bacteria in the culture.

    b. How many bacteria will be in the culture at the end of 9 minutes?

    c. How many bacteria were in the culture initially?

14. Melinda bought a painting in 2000 for $2,500. Since then, the value of the painting increased linearly $300 per year.

    a. Construct a function that represents the value of the painting.

    b. What was the price of the painting in 2007?

    c. What will the price be in 2018?

15. Nathan is renting movies from a local provider. The cost of renting 1 DVD is $3. The cost of renting 2 DVD's is $6. The cost of renting 3 DVD's is $9.

    a. Construct a function that represents the cost of renting DVD's.

    b. How much does it cost to rent 20 DVD's?

16. Felix deposited $500 in a savings account. Each month thereafter, he deposits $200 into the account.

    a. Construct a function that represents the total amount Felix has in his savings account.

    b. How much money does Felix have in his savings account after 2 years?

F-LE.3 Observe using graphs and tables that a quantity increasing exponentially eventually exceeds a quantity increasing linearly or quadratically.

1. Bobby wants to invest $5,000 into a financial plan. Three banks offer the following choices as financial investments:

Bank A: $y = 5000 + 200 \cdot x$

Bank B: $y = 5000 + 50x^2 - 200x$

Bank C: $y = 5000 \cdot (1.02)^x$, where y represents the amount of money after x years .

    a. Which plan is better for Bobby if he invests his money for 1 year? How about 5 years? How about 10 years?

    b. Compare the three financial plans using tables or graphs and find which financial plan is the best for different time periods.

2. If $f(x) = 8x + 3$, $g(x) = x^2 - 2x + 3$ and $h(x) = 3 \cdot (1.36)^x$ are three functions:

    a. What is the Y intercept of each function? What do you observe about the Y intercepts?

    b. Graph the functions f, g and h in the same system of coordinates.

    c. Find the interval where function f is greater than the functions g and h.

    d. Find the interval where function g is greater than the functions f and h.

    e. Find the interval where function h is greater than the functions f and g.

    f. What do you observe about the function h? Is it true that the exponential function exceeds the linear and the quadratic function after a certain value?

3. Jacob wants to invest $9,000 into a financial plan. Three banks offer the following choices as financial investments:

Bank A: $y = 9000 + 450 \cdot x$

Bank B: $y = 9000 + 80x^2 - 300x$

Bank C: $y = 9000 \cdot (1.03)^x$,

where y represents the amount of money after x years.

    a. Which plan is better for Jacob if he invests his money for 2 years? How about 6 years? How about 15 years?

    b. Compare the three financial plans using tables or graphs and find which financial plan is the best for different time periods.

4. If $f(x) = 5x + 17$, $g(x) = \frac{1}{9}x^2 - \frac{2}{3}x + 17$ and $h(x) = 17 \cdot (\frac{11}{18})^x$ are three functions:
   a. What is the Y intercept of each function? What do you observe about the Y intercepts?
   b. Graph the functions f, g and h in the same system of coordinates.
   c. Find the interval where function f is greater than the functions g and h.
   d. Find the interval where function g is greater than the functions f and h.
   e. Find the interval where function h is greater than the functions f and g.
   f. What do you observe about the function h? Is it true that the exponential function exceeds the linear and the quadratic function after a certain value?

5. Elis wants to invest \$8,000 into a financial plan. Three banks offer the following choices as financial investments:

Bank A:	$y = 8000 + 300 \cdot x$
Bank B:	$y = 8000 + 60x^2 - 300x$
Bank C:	$y = 8000 \cdot (1.04)^x$, where y represents the amount of money after x

   years .
   a. Which plan is better for Elis if he invests his money for 1 year? How about 7 years? How about 12 years?
   b. Compare the three financial plans using tables or graphs and find which financial plan is the best for different time periods.

6. If $f(x) = 5x + 12$, $g(x) = x^2 - 4x + 12$ and $h(x) = 12 \cdot (1.28)^x$ are three functions:
   a. What is the Y intercept of each function? What do you observe about the Y intercepts?
   b. Graph the functions f, g and h in the same system of coordinates.
   c. Find the interval where function f is greater than the functions g and h.
   d. Find the interval where function g is greater than the functions f and h.
   e. Find the interval where function h is greater than the functions f and g.
   f. What do you observe about the function h? Is it true that the exponential function exceeds the linear and the quadratic function after a certain value?

_F-LE.5_ _Interpret the parameters in a linear or exponential function in terms of a context._

A-SSE.1 _Interpret expressions that represent a quantity in terms of its context._

   a. Interpret parts of an expression, such as terms, factors, and coefficients.
   b. Interpret complicated expressions by viewing one or more of their parts as a single entity.

1. For the equation $y = m \cdot x + b$ what is the meaning of parameter m and what is the meaning of parameter b?

2. For the equation $y = a \cdot b^x$ what is the meaning of parameter a and what is the meaning of parameter b?

3. Mr. Smith recorded the typing speeds (in words per minute) of 30 students and their weeks of experience. The linear function that describes this experiment is $y = 3.8x + 14.6$, where x is the number of weeks of experience of a student and y is the student's typing speed. What is the meaning of 3.8 in this equation and in the context of this problem? What is the meaning of 14.6 in this equation and in the context of this problem?

4. The equation $y = 350x + 4700$ represents the value (in dollars) of a work of art from 1958 to 2006. What does the number 350 represent in this context? How about 4700?

5. The function $f(x) = -\frac{1}{15}x + 20$ gives the amount of gas $f(x)$ that will be left in the tank of a truck after traveling x miles. Interpret the coefficients $-\frac{1}{15}$ and 20 in the context of this problem.

6. The function $b = 350 - 14m$ shows the balance b in Sofia's checking account after m months. Interpret the coefficients 350 and -14 in the context of this problem.

7. The function $w = 130c + 25000$ represents the total weight w, in pounds, of a truck that carries c cubic feet of bricks. What does 130 represent in this equation and in the context of this problem? What does 25,000 represent in this equation and in the context of this problem?

8. Juan is going to an amusement park. The equation $A = 300 - 12R$ represents the amount of money, A, he has left after R rides. Interpret the coefficients 300 and -12 in the context of this problem.

9. The population of sardines in a lake after x years is represented by the function $f(x) = 4000 - 250x$. What is the meaning of 4000 in this function rule and in the context of this problem? What is the meaning of -250 in this function rule and in the context of this problem?

10. The annual tuition at a private school since 2001 is modeled by the function $f(x) = 7000 \cdot (1.06)^x$, where x is the number of years since 2004. Interpret the coefficient 7,000 and the base of the exponent 1.06 in the context of this problem.

11. The number of bacteria in a culture changes every year according to the function $f(x) = 100 \cdot (1.4)^x$, where x is the number of years since the bacteria appeared. Interpret the coefficient 100 and the base of the exponent 1.4 in the context of this problem.

12. Tania deposited a certain amount of money into a bank. After x years, Tania's amount can be calculated by the function $f(x) = 2000 \cdot (1.005)^{12x}$. Interpret the coefficient 2,000, the base of the exponent 1.005 and the coefficient of the exponent 12, in the context of this problem.

13. DIESEL TAXI cab company charges according to the function $P = 2.3x + 4$, where x is the number of miles traveled. What does the number 2.3 represent in this context? How about 4?

14. The population of a city is modeled by the function $f(x) = 25000 \cdot (1.09)^x$, where x is the number of years since 1945. Interpret the coefficient 25,000 and the base of the exponent 1.09 in the context of this problem.

15. Write a verbal expression for each algebraic expression:

    a. $3 \cdot x^2$

    b. $4 \cdot (a + 9)$

    c. $5x^2 + 3x$

    d. 3^{2x}

    e. $5x + 3$

    f. $x^2 - 3x + 1$

    g. $2 \cdot t$

    h. $\frac{1}{2}p^3$

    i. $(x - y)^2$

    j. $(2x - 3y)^3$

    k. $-x$

    l. $\frac{1}{x}$

    m. $-\frac{1}{x}$

    n. $\frac{a+b}{c-d}$

    o. $x^3 + 25$

16. Write the algebraic expression for each verbal expression:

    a. The sum of a number and 7

    b. 3 more than the triple of a number

    c. Three seventh of the square of a number

d. 10 less a number

e. 8 more than the product of a and b

f. Two times the sum of x and y

g. x more than 9

h. 3 times a number

i. w divided by 14

j. Five times a number plus 7

k. Five times the sum of a number and 7

l. b squared minus 2

m. a number cubed

n. The difference between one third of a number and 10

o. The quotient of x and 8

p. The square of the quotient of 3 and b

q. 25 decreased by 7 times x

r. The opposite of x

s. The reciprocal of $\frac{a}{b}$

t. The opposite of the reciprocal of a number

17. Luke earns $9 per hour working at a restaurant and $18 for each car he washes. Write an algebraic expression to show the amount of money Luke earns working h hours at the restaurant and washing c cars.

18. Emilia has in her pocket q quarters, d dimes, n nickels and p pennies. Write an algebraic expression that represents the amount of money which Emilia has in her pocket. Using this algebraic expression, find the amount of money that Emilia has, if there are 9 quarters, 4 dimes, 6 nickels and 12 pennies in her pocket.

19. The function $r(x) = 60x$ represents the number of words $r(x)$ you can type in x minutes. What does the coefficient 60 represent in the context of this problem?

20. The equation $d = 120000t$ gives the distance d, in miles, that sound travels in t seconds. What does the coefficient 120000 represent in the context of this problem?

21. The function $d(x) = 14x$ represents the distance $d(x)$, in miles, that the truck can travel with x gallons of gasoline. What does the coefficient 14 represent in the context of this problem?

22. Suppose a ball is dropped from a certain height. The function $h(n) = 7 \cdot (0.87)^n$, represents the maximum height of the ball after the nth bounce.

a. What does the coefficient 7 represent in the context of this problem?

b. What does the factor 0.87 represent in the context of this problem?

23. The value of Mr. Johnson's car, x years after it is purchased is given by the function $V(x) = 14,000(0.91)^x$.

    a. What does the coefficient 14,000 represent in the context of this problem?

    b. What does the factor 0.91 represent in the context of this problem?

24. The function $p(t) = 2.7 \cdot (1.08)^t$ gives a city's population $p(t)$ (in millions), where t is the number of years since 1998.

    a. What does the coefficient 2.7 represent in the context of this problem?

    b. What does the factor 1.08 represent in the context of this problem?

25. The annual tuition at a college since 1980 is modeled by the equation $T = 4000(1.05)^x$, where T is the tuition cost and x is the number of years 1980.

    a. What does the coefficient 4000 represent in the context of this problem?

    b. What does the factor 1.05 represent in the context of this problem?

26. A bank employee notices an abandoned checking account with a certain balance. The bank charges a certain monthly fee for the account. The function $b = 240 - 5m$ shows the balance b in the account after m month.

    a. What does the coefficient -5 represent in the context of this problem?

    b. What does the term 240 represent in the context of this problem?

27. The equation $w = 125c + 20000$ represents the total weight w, in pounds, of a truck that carries c cubic feet of bricks.

    a. What does the coefficient 125 represent in the context of this problem?

    b. What does the term 20000 represent in the context of this problem?

28. The cost C, in dollars, for delivered sandwiches depends on the number S of sandwiches ordered. This situation is represented by the equation: $C = 5 + 9S$.

    a. What does the coefficient 9 represent in the context of this problem?

    b. What does the term 5 represent in the context of this problem?

29. Felix, Emilia and Nancy are going to an amusement park. The function $A = 200 - 15R$ represents the amount of money, A, they have left after R rides.

    a. What does the coefficient -15 represent in the context of this problem?

    b. What does the term 200 represent in the context of this problem?

G-CO.1 *Know precise definitions of angle, circle, perpendicular line, parallel line, and line segment, based on the undefined notions of point, line and distance along a line.*

1. Define and draw an angle.
2. How do you measure an angle? What is the unit of measure for an angle?
3. Draw:
    a. An angle that has 30°.
    b. An angle that has 60°.
    c. An angle that has 90°.
    d. An angle that has 120°.
    e. An angle that has 150°.
    f. An angle that has 180°.
4. What is an *acute* angle? What is an *obtuse* angle? What is a *right* angle?
5. Define and draw a circle. What are the elements of a circle?
6. Draw a circle with the radius of 2 inches.
7. Give examples of circles in real-life situations.
8. Define and draw two perpendicular lines.
9. Give examples of two perpendicular lines in real-life situations.
10. Define and draw two parallel lines.
11. Give examples of two parallel lines in real-life situations.
12. What is the difference between a line and a line segment?
13. Draw a line segment and find the length of it.
14. Draw a line segment with the length of 5 inches.

G-GPE.4 *Use coordinates to prove simple geometric theorems algebraically. For example, prove or disprove that a figure defined by four given points in the coordinate plane is a rectangle; prove or disprove that the point $(1, \sqrt{3})$ lies on the circle centered at the origin and containing the point (0, 2).*

G-GPE.5 *Prove the slope criteria for parallel and perpendicular lines and use them to solve geometric problems (for example, find the equation of a line parallel or perpendicular to a given line that passes through a given point).*

1. If $A(x_1, y_1)$ and $B(x_2, y_2)$ are two given points, what is the distance between A and B? What is the length of segment AB?
2. Find the distance between the two given points:
   a. $A(3,7)$ and $B(5,6)$
   b. $C(-3,7)$ and $D(5,6)$
   c. $E(3,-7)$ and $F(-5,6)$
   d. $G(-3,-7)$ and $H(5,-6)$
   e. $I(5,-7)$ and $J(5,9)$
   f. $K(-8,10)$ and $L(3,10)$
   g. $M(0,-1)$ and $N(-1,0)$
   h. $P(2,0)$ and $Q(-3,0)$
3. If $A(x_1, y_1)$ and $B(x_2, y_2)$ are two given points, what is the slope of line AB?
4. If $ax + by = c$ is the equation of a line, what is the slope of this line?
5. If $y = m \cdot x + b$ is the equation of a line, what is the slope of this line?
6. If two lines are *parallel*, what is the relationship between their slopes?
7. If two lines are *perpendicular*, what is the relationship between their slopes?
8. Find the slope of the line AB, knowing the coordinates of A and B: $A(3,5)$ and $B(5,13)$.
9. Find the slope of the line PQ, knowing the coordinates of P and Q: $P(-2,7)$ and $Q(8,-4)$.
10. Find the slope of the line MN, knowing the coordinates of M and N: $M(0,-4)$ and $N(8,-7)$.
11. Find the slope of the line AB, knowing the coordinates of A and B: $A(-1,2)$ and $B(-4,-13)$.
12. Find the slope of the line PQ, knowing the coordinates of P and Q: $P(1,-7)$ and $Q(5,-7)$.
13. Find the slope of the line JK, knowing the coordinates of J and K: $J(9,-2)$ and $K(9,3)$.
14. If $2x + y = 6$ is the equation of a line, find the coordinates of a point on this line.
15. If $x - 3y + 7 = 0$ is the equation of a line, find the coordinates of a point on this line.
16. If $y = 4x - 9$ is the equation of a line, find the coordinates of a point on this line.
17. If $3x - 4y = 12$ is the equation of a line:
    a. Find the coordinates of two different points on this line and plot these points.
    b. Using the coordinates of the two points found at point (a), find the slope of this line.

18. If $x + 5y - 10 = 0$ is the equation of a line:
  a. Find the coordinates of two different points on this line and plot these points.
  b. Using the coordinates of the two points found at point (a), find the slope of this line.

19. If $y = 7x - 13$ is the equation of a line:
  a. Find the coordinates of two different points on this line and plot these points.
  b. Using the coordinates of the two points found at point (a), find the slope of this line.

20. If $y = 4$ is the equation of a line:
  a. Find the coordinates of two different points on this line and plot these points.
  b. Using the coordinates of the two points found at point (a), find the slope of this line.

21. If $4x + 5y = 10$ is the equation of a line:
  a. Find the coordinates of two different points on this line and plot these points.
  b. Using the coordinates of the two points found at point (a), find the slope of this line.

22. If $2x - 3y + 18 = 0$ is the equation of a line:
  a. Find the coordinates of two different points on this line and plot these points.
  b. Using the coordinates of the two points found at point (a), find the slope of this line.

23. If $x = 4$ is the equation of a line:
  a. Find the coordinates of two different points on this line and plot these points.
  b. Using the coordinates of the two points found at point (a), find the slope of this line.

24. If $y = -5x + 3$ is the equation of a line:
  a. Find the coordinates of two different points on this line and plot these points.
  b. Using the coordinates of the two points found at point (a), find the slope of this line.

25. Find the slope of the lines given by the equation:
  a. $y = 4x + 7$
  b. $3x - 5y = 14$
  c. $x + 3y - 10 = 0$
  d. $y = \frac{3}{7}x - 8$
  e. $3y = 6x - 10$
  f. $x = -2$
  g. $4x + y = 8$
  h. $-3x + 2y + 7 = 0$
  i. $y = 3$
  j. $x = 3y - 7$
  k. $4y - 2x = 9$
  l. $3y + 6x - 8 = 0$
  m. $y = -5$

26. Using the *slope criteria*, tell whether each pair of lines is *parallel, perpendicular* or *neither*:

    a. $y = 3x + 5$ and $y = 3x - 10$

    b. $y = 4x - 10$ and $y = -\frac{1}{4}x + 5$

    c. $y = 7x + 1$ and $y = -7x - 3$

    d. $2x + 3y = 7$ and $4x + 6y = -19$

    e. $x - 3y + 5 = 0$ and $-2x + 6y + 13 = 0$

    f. $2y = 4x - 10$ and $12y = -6x + 5$

    g. $4x - y = 10$ and $4y = -x + 12$

    h. $y = -\frac{1}{3}x + 8$ and $2x + 6y = 17$

    i. $y = 4$ and $y = -3$

    j. $y = 7$ and $x = 4$

    k. $x = -2$ and $x = 1$

    l. $y = 2x - 5$ and $y = 2$

27. Write an equation for the line that is *parallel* to the given line and that passes through the given point:

    a. $y = 4x + 5$, $A(3,7)$

    b. $y = -2x + 7$, $B(-1,4)$

    c. $2x - 3y = 9$, $C(4,-2)$

    d. $x + 4y - 5 = 0$, $D(-5,6)$

    e. $y = 3$, $E(7,-4)$

    f. $x = -4$, $F(-1,6)$

28. Write an equation for the line that is *perpendicular* to the given line and that passes through the given point:

    a. $y = 4x + 5$, $A(3,7)$

    b. $y = -2x + 7$, $B(-1,4)$

    c. $2x - 3y = 9$, $C(4,-2)$

    d. $x + 4y - 5 = 0$, $D(-5,6)$

    e. $y = 3$, $E(7,-4)$

    f. $x = -4$, $F(-1,6)$

29. Write an equation of the line that is *parallel* to the Y-axis and that is 4 units to the right of the Y-axis.

30. Write an equation of the line that is *perpendicular* to the Y-axis and that is 3 units below the X-axis.

31. Write an equation of the line that is *parallel* to the X-axis and that is 5 units above the X-axis.

32. Write an equation of the line that is *perpendicular* to the X-axis and that is 2 units to the left of the Y-axis.

33. Find the value of parameter a knowing that the line $y = ax + 7$ is *parallel* to the line $y = 4x - 10$.

34. Find the value of parameter m knowing that the line $y = 4x - 3$ is *perpendicular* to the line $y = (m + 1)x - 2$.

35. On a map of downtown, 6th street is *perpendicular* to Avenue G. The equation $y = 4x - 7$ represents 6th street. What is the equation representing Avenue G if it passes through the point (3, 14)?

36. At the airport, the new runway will be *parallel* to a nearby highway. On the scale drawing of the airport, the equation that represents the highway is $2y = 6x - 11$. What is the equation of the new runway if it passes through the point (7, 2)?

37. Check if the triangle with the vertices $A(4, 2)$, $B(7, -1)$ and $C(11,3)$ is a right angle triangle. Explain your reasoning.

38. Check if the triangle with the vertices $A(1,3)$, $B(3,1)$ and $C(5,5)$ is an isosceles triangle. Explain your reasoning.

39. Check if the point $(1, \sqrt{3})$ lies on the circle centered at the origin and containing $(0,2)$. Explain your reasoning.

40. Check if the quadrilateral with the vertices $A(3,1)$, $B(5, -3)$, $C(9, -1)$ and $D(7,3)$ is a rectangle. Explain your reasoning.

41. Construct a rectangle by finding the coordinates of the four vertices.

G-GPE.6 *Find the point on a directed line segment between two given points that partitions the segment in a given ratio.*

Note: *At this level, focus on finding the midpoint of a segment.*

G-GPE.7 *Use coordinates to compute perimeters of polygons and areas of triangles and rectangles, for example, using the distance formula.*

1. Find the coordinates of the midpoint of segment AB, if $A(4,9)$ and $B(10,3)$.
2. Find the coordinates of the midpoint of segment CD, if $C(4,-6)$ and $D(-10,2)$.
3. Find the coordinates of the midpoint of segment EF, if $E(-1,7)$ and $F(1,-5)$.
4. Find the coordinates of the midpoint of segment GH, if $G(8,-2)$ and $H(-4,-9)$.
5. Find the coordinates of the midpoint of segment JK, if $J(-14,-18)$ and $K(16,24)$.
6. Find the coordinates of the midpoint of segment LM, if $L(0,-1)$ and $M(0,5)$.
7. On a map, Louisville is located at $(4,-3)$ and Charlesville is located at $(-7,1)$. What is the distance between Louisville and Charlesville? If Ducksburg is halfway between Louisville and Charlesville what is the distance between Louisville and Ducksburg?
8. If $M(-3,4)$ is the midpoint of segment PQ and $P(7,-1)$, what are the coordinates of Q?
9. On a map, Sofia's house is located at $(-9,1)$ and Tatiana's house is located at $(5,-8)$. What are the coordinates for Diana's house if she lives exactly halfway between Sofia and Tatiana? What is the distance between Sofia's house and Diana's house?
10. The midpoint of XY is point $M(-10,3)$. If the coordinates of X are $(-6,7)$, what are the coordinates of Y?
11. Find the perimeter of the triangle ΔMNP with the vertices $M(2,6), N(7,-1)$ and $P(-5,-3)$.
12. Find the perimeter of the quadrilateral $PQRS$ with the vertices $P(-2,3), Q(1,-5),$ $R(4,7)$ and $S(-1,-9)$.
13. Find the area of the rectangle $ABCD$ with the vertices $A(3,1), B(5,-3), C(9,-1)$ and $D(7,3)$.
14. Find the area of the right angle triangle ΔABC with the vertices $A(4,2), B(7,-1)$ and $C(11,3)$.
15. Find the perimeter of the quadrilateral $PQRS$ with the vertices $P(1,4), Q(7,-4),$ $R(0,-10)$ and $S(-6,2)$.
16. Find the perimeter of the triangle ΔABC with the vertices $A(-12,5), B(9,-3)$ and $C(1,8)$.
17. Find the area of the triangle ΔABC with the vertices $A(-8,5), B(7,-2)$ and $C(5,10)$.

G-GMD.1 *Give an informal argument for the formulas for the circumference of a circle, area of a circle, volume of a cylinder, pyramid, and cone. Use dissection arguments, Cavalieri's principle, and informal limit arguments.*

Note: *Informal limit arguments are not the intent at this level.*

G-GMD.3 *Use volume formulas for cylinders, pyramids, cones, and spheres to solve problems.*

Note: *At this level, formulas for pyramids, cones and spheres will be given.*

The Circumference of a Circle $= 2\pi R$, where R is the radius

The Area of a Circle $= \pi R^2$, where R is the radius

The Volume of a Pyramid $= \frac{B \cdot h}{3}$, where B is the base area and h is the height

The Volume of a Cylinder $= \pi R^2 \cdot h$, where R is the radius and h is the height

The Volume of a Cone $= \frac{\pi R^2 \cdot h}{3}$, where R is the radius and h is the height

The Volume of a Sphere $= \frac{4\pi R^3}{3}$, where R is the radius

The Surface Area of a Sphere $= 4\pi R^2$, where R is the radius

---

1. Give an informal argument for the formula for the circumference of a circle.
2. Find the circumference of a circle with the radius 5 inches.
3. Find the radius of a circle with a circumference of 18 inches.
4. Give an informal argument for the formula for the area of a circle.
5. Find the area of a circle with the radius of 4 inches.
6. Give an informal argument for the formula for the volume of a cylinder.
7. Find the volume of a cylinder with the height of 12 centimeters and radius of 3 centimeters.
8. Give an informal argument for the formula for the volume of a cone.
9. Find the volume of a cone with the height of 24 centimeters and radius of 8 centimeters.
10. Give an informal argument for the formula for the volume of a pyramid.
11. Find the volume of a pyramid that has a height of 5 feet and a square base with the length of the side equal to 8 feet.
12. Find the volume of a pyramid that has a height of 10 yards and an equilateral triangle base with the length of the side equal to 7 yards.

13. Find the volume of a pyramid that has a height of 8 inches and a square base with sides 3 inches long.

14. The volume of a square pyramid is 738 centimeters cubed. If the height is 6 centimeters, then find the dimensions of the base.

15. Mr. Johnson wants to build a square pyramid out of cement. He wants the pyramid to be 8 feet tall and its base edge to be 12 feet long. How many cubic feet of cement will he need?

16. Elis made a square pyramid-shaped candle. The volume of the candle is 1183 cubic centimeters and its base has an area of 169 square centimeters. How high is the candle?

17. Find the volume of a pyramid that has a height of 7 meters and an equilateral triangle base with the length of the side equal to 8 meters.

18. Gabby has a pyramid-shaped precious stone that has a height of 1 inch and an equilateral triangle base with sides 1 inch long as well. Find the volume of Gabby's precious stone.

19. Tania has a cone-shaped party hat. The diameter of the base of the hat is 10 inches long and the height is 7 inches. Tania wants to fill her party hat with confetti. How much will the confetti cost if 2 cubic inches of confetti cost $1.25?

20. Sharon is packing cylindrical cans with diameter of 10 centimeters and height of 16 centimeters into a box with the length of 90 centimeters, the width of 40 centimeters and the height of 48 centimeters. All rows must contain the same number of cans. The cans must touch each other. When there is no more space for cans, Sharon fills all the empty space in the box with packing foam.

    a. How many cans could Sharon pack in the box?

    b. Find the volume of packing foam she uses.

    c. What percentage of the box's volume is filled by the foam?

21. The circumference of a great circle of a sphere is 48.18π feet. What is the volume of the sphere?

22. In order to make an iron ball, a cylindrical iron tube is heated and pressed into a spherical shape with the same volume. The height of the cylindrical tube is 3 inches and the radius is 1 inch. Find the radius of the iron ball.

23. The diameter of a baseball is 2.9 inches. Estimate the amount of leather used to cover the baseball.

24. The diameter of Planet Earth is 12756.2 kilometers (7926.3 miles).

    a. What is the surface area of Planet Earth?

    b. What is the volume of Planet Earth?

25. If the radius of a sphere doubles, does its volume double also? Explain your reasoning.

26. The dome of a cathedral has a cone form with the radius of 20 meters and the height of 8 meters. What is the volume of the dome?

27. In a conical tank, the depth and radius of the water level is 14.5 inches and 8 inches respectively. Find the volume of water in the tank.

S-ID.1 Represent data with plots on the real number line (dot plots, histograms, and box plots).

S-ID.2 Use statistics appropriate to the shape of the data distribution to compare center (median, mean) and spread (interquartile range, standard deviation) of two or more different data sets.

S-ID.3 Interpret differences in shape, center, and spread in the context of the data sets, accounting for possible effects of extreme data points (outliers).

1. Create a *histogram*, a *box-plot* and a *stem-plot* for the following data that represents the Algebra 1 test scores of 35 students:
   28, 37, 63, 66, 67, 68, 68, 69, 72, 75, 77, 81, 82, 86, 88, 88, 88, 89, 89, 89, 90, 90, 90, 91, 93, 93, 94, 95, 95, 96, 97, 100, 100, 100, 100
   Describe the shape, center and spread of this data.
2. What are the elements of a *box-plot*?
3. The data below shows the number of people living in a house for 15 residences in a small neighborhood of a city:
   2, 1, 5, 1, 3, 4, 3, 2, 4, 2, 4, 2, 3, 4, 3
   Create a *dot plot* to display the given data and then describe the shape, center and spread of this data.
4. Which of the following types of data are *categorical* and which are *quantitative*:
   a. Heights of students in a classroom
   b. Colors of cars in a parking lot
   c. Ages of professional football players in North Carolina
   d. Test scores
   e. Ice-cream flavors
   f. Speeds of different cars on a high-way
   g. Courses taken by a student
5. The data below shows the number of home runs in a baseball tournament:
   5, 13, 16, 4, 17, 14, 5, 12, 3, 10, 4, 5, 10, 7
   a. Represent this data using a *histogram*.
   b. Represent this data using a *box-plot*.
   c. Describe the shape, center and spread of this data.
6. The data below shows the number of people entering a store during a day for different time intervals:
   40, 45, 53, 72, 45, 50, 67, 75, 82, 98, 62
   a. Represent this data using a *histogram*.
   b. Represent this data using a *box-plot*.
   c. Describe the shape, center and spread of this data.

7. The data below shows the number of hours per week spent watching television by a certain group of people:

14, 10, 13, 4, 9, 16, 23, 7, 9, 1, 11, 14, 17, 14, 24, 15, 13

    a. Represent this data using a *histogram.*

    b. Represent this data using a *box-plot.*

    c. Describe the shape, center and spread of this data.

8. Describe the shape of the following distributions. Use words like: symmetric, bell-shaped, skewed to the left, skewed to the right, uniform, unimodal, bimodal, gap, outlier.

    a.

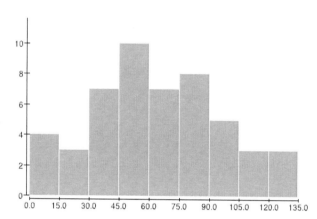

    b.

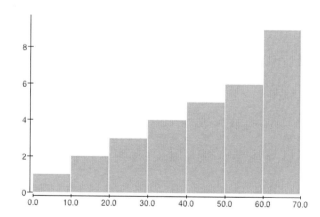

c.

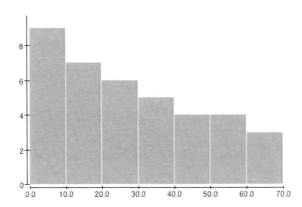

d.

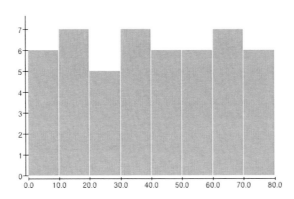

e.

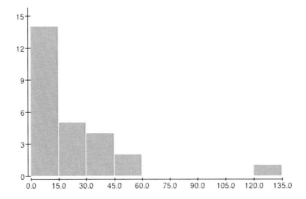

f.

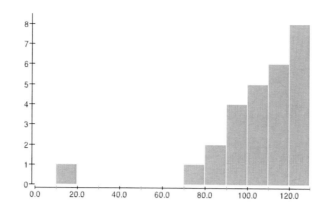

g.

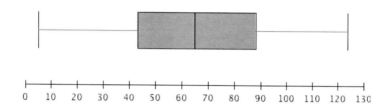

h.

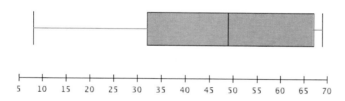

i.

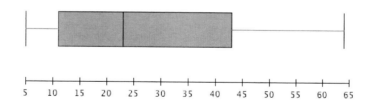

j.

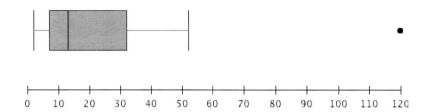

k.

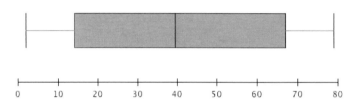

l.

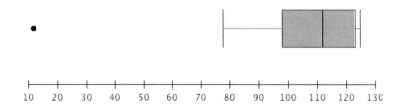

9. Interpret the data represented by the following box-plot:

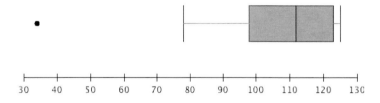

10. The data below shows the amounts in dollars that Felix spent for gas over the last year:
160, 140, 135, 120, 123, 115, 430, 240, 132, 138, 141, 152
- **a.** Find the *mean* and the *median* of this data.
- **b.** Find the *interquartile range* and the *standard deviation* of this data.
- **c.** Represent this data using a *histogram.*
- **d.** Represent this data using a *box-plot.*
- **e.** Describe the center and spread of the data using statistics appropriate to the shape of the data.

11. What does the *Standard Deviation* show about a data?

12. The average monthly income in millions of dollars at a Shopping Mall is shown in the table below:

Jan	Feb	Mar	Apr	May	Jun	Jul	Aug	Sep	Oct	Nov	Dec
2.3	1.9	2.1	1.8	2.2	2.2	2.0	2.8	2.2	2.1	2.5	7.9

    a. Represent this data using a histogram and a box-plot.

    b. According to the shape of this distribution, what measure of spread is more appropriate for this data (*Standard Deviation* or *Interquartile Range*)?

    c. According to the shape of this distribution, what measure of center is more appropriate for this data (*Mean* or *Median*)?

    d. Are there any *outliers*? Explain your reasoning.

13. At a grocery store, Mr. Smith recorded the number of apples that 10 customers bought:
7, 4, 5, 6, 7, 6, 5, 7, 9, 10, 18, 30

    a. Represent this data using a histogram and a box-plot.

    b. According to the shape of this distribution, what measure of spread is more appropriate for this data (*Standard Deviation* or *Interquartile Range*)?

    c. According to the shape of this distribution, what measure of center is more appropriate for this data (*Mean* or *Median*)?

    d. Are there any *outliers*? Explain your reasoning.

14. The salaries (in millions of dollars) of NFL players can be summarized by the *five-number summary* below:

Minimum	Q_1	Median	Q_3	Maximum
323	2,375	5,667	12,678	26,000

    a. Would the maximum or minimum salaries be considered *outliers*?

    b. Construct a box-plot for this distribution and describe the shape of this distribution.

15. Define *percentiles*.

16. Define *Standardized scores* or *Z-scores*.

17. You scored an 87 on a test in your Algebra 1 class where the mean was 85 and the standard deviation was 3. Your best friend is in a different Algebra 1 class and scored a 90 where the mean in her class was 88 and the standard deviation was 4. Who had the better score relative to their own class?

18. We know that the mean weight of eight athletes is 160 pounds. We have the list of weights for seven of the athletes but have lost the weight for one of them. What is the weight of the 8th athlete? The weights of the seven athletes are:
143, 152, 171, 180, 162, 159, 184

19. If the mean of a data representing test scores is currently 83 and the standard deviation is 2.7, what would the new mean and standard deviation be if each individual's score is increased by 1.2?

20. Which of the following distributions will have the smallest standard deviation, assuming that none contains outliers?
   a. A uniform distribution of integers with a mean of 10 and a range of 20.
   b. A bell-shaped distribution of integers with a mean of 10 and a range of 20.
   c. A right-skewed distribution of integers with a mean of 10 and a range of 20.

21. Consider the length of time (in minutes) it takes 20 people to get to the downtown of a city by walking: 15, 13, 17, 16, 35, 27, 5, 16, 18, 21, 5, 37, 12, 25, 7, 23, 50, 2, 36, 59.
   a. One person takes 59 minutes to get to school. Would you consider this an outlier?
   b. Construct a box-plot and a histogram for this set of data.
   c. Describe the distribution.

22. The data below shows the customer ratings for 2 hotel resorts.
Hotel A: 4.3, 4.1, 3.7, 4.5, 3.2, 4.2, 4.3, 3.8, 4.0, 3.1, 4.9, 3.8, 3.2, 3.6, 4.1
Hotel B: 4.4, 2.5, 3.9, 3.7, 3.8, 2.9, 4.3, 3.6, 4.1, 2.8, 4.9, 3.4, 3.8, 4.2, 3.9
   a. Construct parallel box-plots of the two distributions.
   b. Create a histogram for each distribution.
   c. Compare center and spread of the two data sets using statistics appropriate to the shape of the distributions.

23. Which of the following four sets of data have the smallest and largest standard deviations?

I	II	III	IV
1	1	1	1
2	5	1	3
3	5	1	3
4	5	1	3
5	5	5	5
6	5	9	7
7	5	9	7
8	5	9	7
9	9	9	9

24. The game attendance for the 20 home games and the 20 away games of the season for a football team in England are shown below.

Home games	Away games
33,000	30,000
34,000	30,600
34,400	33,200
36,900	34,600
37,500	34,800
40,100	35,100
41,200	36,000
44,500	39,900
46,200	40,000
46,500	40,200
46,600	41,700
49,100	43,800
51,200	44,000
53,200	44,400
53,400	45,000
53,700	47,000
54,100	47,600
54,400	48,000
54,700	51,700
55,900	54,300

    a. Construct parallel box-plots of the two distributions.

    b. Create a histogram for each distribution.

    c. Compare center and spread of the two data sets using statistics appropriate to the shape of the distributions.

25. The data below shows the distance traveled in miles during a year by two different families.

The Smith's: 450, 370, 430, 420, 410, 480, 850, 580, 470, 350, 310, 430

The Johnson's: 480, 360, 425, 610, 490, 395, 1250, 440, 230, 250, 360, 480

    a. Construct parallel box-plots of the two distributions.

    b. Create a histogram for each distribution.

    c. Compare center and spread of the two data sets using statistics appropriate to the shape of the distributions.

26. The data below shows the number of extra math problems done by different students in a certain class:

4, 7, 1, 5, 3, 0, 4, 2, 10, 9, 1, 12, 0, 5, 8

    a. Find the *mean* and the *median* of this data.

    b. Find the *interquartile range* and the *standard deviation* of this data.

    c. Represent this data using a *histogram.*

    d. Represent this data using a *box-plot.*

    e. Describe the center and spread of the data using statistics appropriate to the shape of the data.

27. The data below shows the ages of Felix's friends:

29, 32, 27, 31, 34, 28, 29, 20, 22, 45, 53, 18, 14, 12, 34, 36, 32, 33, 25, 23, 32, 37, 35, 35, 33, 34, 29

    a. Represent this data using a histogram and a box-plot.

    b. According to the shape of this distribution, what measure of spread is more appropriate for this data (*Standard Deviation* or *Interquartile Range*)?

    c. According to the shape of this distribution, what measure of center is more appropriate for this data (*Mean* or *Median*)?

    d. Are there any *outliers*? Explain your reasoning.

28. The data below shows the heights of the buildings (in meters) in a certain city:

37, 83, 92, 81, 52, 10, 75, 93, 84, 92, 87, 43, 52, 61, 73

    a. Represent this data using a histogram and a box-plot.

    b. According to the shape of this distribution, what measure of spread is more appropriate for this data (*Standard Deviation* or *Interquartile Range*)?

    c. According to the shape of this distribution, what measure of center is more appropriate for this data (*Mean* or *Median*)?

    d. Are there any *outliers*? Explain your reasoning.

29. The data below shows the Final Algebra 1 Exam scores of 24 students:

67, 78, 93, 82, 67, 89, 31, 81, 54, 78, 71, 84, 81, 79, 94, 92, 73, 100, 85, 80, 82, 91, 68, 70

    a. Represent this data using a histogram and a box-plot.

    b. According to the shape of this distribution, what measure of spread is more appropriate for this data (*Standard Deviation* or *Interquartile Range*)?

    c. According to the shape of this distribution, what measure of center is more appropriate for this data (*Mean* or *Median*)?

    d. Are there any *outliers*? Explain your reasoning.

30. The data below shows the ages of people attending two movies.

Movie A: 22, 24, 23, 21, 24, 22, 25, 28, 31, 32, 26, 25, 27, 35

Movie B: 34, 31, 45, 47, 50, 52, 63, 25, 18, 47, 50, 71, 65, 59

    a. Construct parallel box-plots of the two distributions.

    b. Create a histogram for each distribution.

    c. Compare center and spread of the two data sets using statistics appropriate to the shape of the distributions.

31. The data below shows the average car speeds (in miles/hour) recorded by a radar during a day on two highways in North Carolina.

Highway 1: 40, 42, 48, 57, 55, 41, 44, 41, 42, 44, 49, 52, 55, 58, 61

Highway 2: 43, 45, 46, 61, 52, 47, 44, 40, 45, 52, 62, 53, 42, 53, 64

    a. Construct parallel box-plots of the two distributions.

    b. Create a histogram for each distribution.

    c. Compare center and spread of the two data sets using statistics appropriate to the shape of the distributions.

32. Compare center and spread of the two distributions given:

    a.

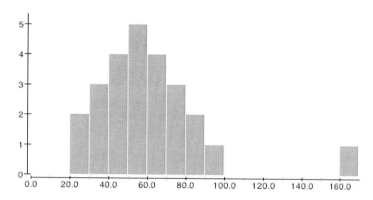

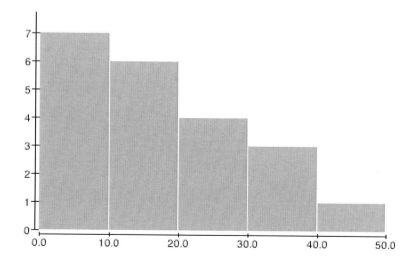

b.

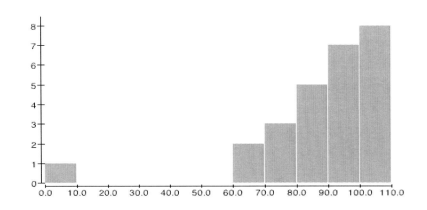

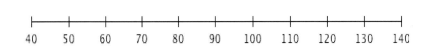

c.

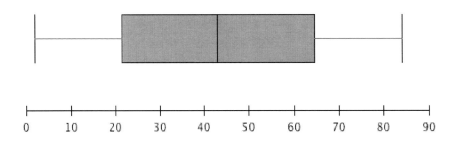

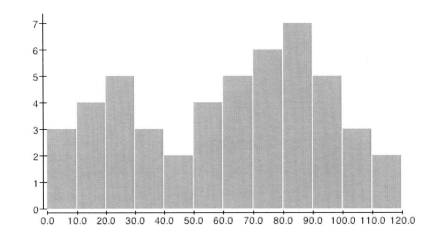

d.

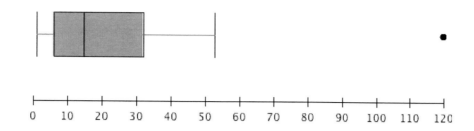

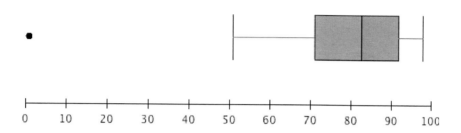

33. The table below shows the average annual sales (in millions of dollars) at a Shopping Center for the last 10 years.

Year	2002	2003	2004	2005	2006	2007	2008	2009	2010	2011
Sales (in millions)	9.3	9.8	8.7	15.9	7.2	6.9	5.8	7.5	5.4	7.0

    a. Find the mean and the standard deviation of the data.

    b. In what years were the sales less than one standard deviation from the mean?

    c. Find the median and interquartile range for the data.

S-ID.5 *Summarize categorical data for two categories in two-way frequency tables. Interpret relative frequencies in the context of the data (including joint, marginal, and conditional relative frequencies). Recognize possible associations and trends in the data.*

1. A survey of 20 adults asked: "Of SEDAN, SUV and TRUCK, which is your favorite type of car?" The results were as follows:

SEDAN	SUV	TRUCK	SUV
SUV	SEDAN	TRUCK	SUV
TRUCK	SUV	SUV	TRUCK
SUV	SEDAN	TRUCK	SUV
TRUCK	SUV	SEDAN	SUV

    a. Create a *frequency table* that summarizes these data.

    b. Create a *relative frequency table* of the same data.

2. A survey of 20 adults asked: "Of SEDAN, SUV and TRUCK, which is your favorite type of car?" Each respondent's gender was also recorded. The results were as follows:

Type of car	Gender
SEDAN	FEMALE
SUV	FEMALE
TRUCK	MALE
SUV	FEMALE
SUV	MALE
SEDAN	FEMALE
TRUCK	MALE
SUV	FEMALE
TRUCK	FEMALE
SUV	FEMALE
SUV	MALE
TRUCK	MALE
SUV	FEMALE
SEDAN	FEMALE
TRUCK	MALE
SUV	FEMALE
TRUCK	MALE
SUV	FEMALE
SEDAN	MALE
SUV	FEMALE

    a. Create a *two-way frequency table* that summarizes these data including *marginal frequencies*. Interpret the *joint frequencies* of this table.

    b. Create a *two-way relative frequency table* that summarizes these data which displays *conditional frequencies* for **gender**. Interpret the *conditional frequencies* of this table.

    c. Create a *two-way relative frequency table* that summarizes these data which displays *conditional frequencies* for **type of car**. Interpret the *conditional frequencies* of this table.

3. A survey of 12 children asked: "Of RED, BLUE, GREEN and ORANGE, which is your favorite color?" Each child's gender was also recorded. The results were as follows:

Color	Gender
ORANGE	FEMALE
BLUE	MALE
ORANGE	FEMALE
RED	FEMALE
BLUE	FEMALE
ORANGE	FEMALE
GREEN	MALE
BLUE	MALE
ORANGE	FEMALE
RED	FEMALE
ORANGE	FEMALE
BLUE	MALE

   a. Create a *two-way frequency table* that summarizes these data including *marginal frequencies*. Interpret the *joint frequencies* of this table.

   b. Create a *two-way relative frequency table* that summarizes these data which displays *conditional frequencies* for **gender**. Interpret the *conditional frequencies* of this table.

   c. Create a *two-way relative frequency table* that summarizes these data which displays *conditional frequencies* for **color**. Interpret the *conditional frequencies* of this table.

4. A survey of 18 people asked: "Of SPRING, SUMMER, FALL and WINTER, which is your favorite season of the year?" Each person's occupation was also recorded. The results were as follows:

Season	Occupation
SPRING	ARCHITECT
SUMMER	TEACHER
WINTER	PHARMACIST
SPRING	ARCHITECT
FALL	PHARMACIST
SUMMER	TEACHER
SUMMER	TEACHER
SPRING	ARCHITECT
WINTER	TEACHER
SUMMER	ARCHITECT
FALL	PHARMACIST
SPRING	PHARMACIST
FALL	PHARMACIST
SUMMER	TEACHER
WINTER	ARCHITECT
FALL	PHARMACIST
SUMMER	TEACHER
FALL	ARCHITECT

a. Create a *two-way frequency table* that summarizes these data including *marginal frequencies.* Interpret the *joint frequencies* of this table.
b. Create a *two-way relative frequency table* that summarizes these data which displays *conditional frequencies* for **occupation**. Interpret the *conditional frequencies* of this table.
c. Create a *two-way relative frequency table* that summarizes these data which displays *conditional frequencies* for **season**. Interpret the *conditional frequencies* of this table.

S-ID.6 *Represent data on two quantitative variables on a scatter plot, and describe how the variables are related.*

   a. *Fit a function to the data; use functions fitted to data to solve problems in the context of the data. Use given functions or choose a function suggested by the context. Emphasize linear and exponential models.*

   b. *Informally assess the fit of a function by plotting and analyzing residuals.*
   Note: *At this level, for part b, focus on linear models.*

   c. *Fit a linear function for a scatter plot that suggests a linear association.*
S-ID.7 *Interpret the slope (rate of change) and the intercept (constant term) of a linear model in the context of the data.*

S-ID.8 *Compute (using technology) and interpret the correlation coefficient of a linear fit.*

S-ID.9 *Distinguish between correlation and causation.*

1. The data below shows the number of hours spent studying for the Algebra 1 final exam and the final exam grade of 12 students.

Number of hours	0	1	2	3	4	5	6	7	8	9	10	11
Final Exam grade	20	38	54	58	64	71	73	76	81	88	91	94

   a. Represent the data on a scatter plot and describe how the variables are related using the shape, strength and direction of the scatter plot.
   b. Find the linear function that best fits this data. What is the meaning of the slope and the Y-intercept of this linear function in the context of this data?
   c. According to the line of best fit, what would be the final exam grade of a student who studies 8 and a half hours for this exam?
   d. According to the line of best fit, how many hours should a student study in order to get a 100 on the final exam?
   e. Calculate the residuals from the plot above. What do they represent? Are the points with positive residuals located above or below the regression line?
   f. What is the sum of the squared residuals of the linear model that represents the situation described above? Is there any different line that gives a smaller sum? Explain your reasoning.
   g. What is the Least Squares Regression line that models this data? How do you know if this equation is the line of best fit to model the data?
   h. Using technology, compute the correlation coefficient and interpret what it means in the context of the data. (What does the correlation coefficient measure for a linear association?)
   i. Analyze correlation and causation for this data set.

2. The data below shows the height (in feet) and the shoe sizes (in inches) of ten people.

Height	5.4	5.6	5.7	5.8	6	6.1	6.3	6.4	6.8	7.1
Shoe size	9.2	9.7	9.5	9.3	10.4	11.2	12.5	12.2	12.9	13.6

    a. Represent the data on a scatter plot and describe how the variables are related using the shape, strength and direction of the scatter plot.

    b. Find the linear function that best fits this data. What is the meaning of the slope and the Y-intercept of this linear function in the context of this data?

    c. According to the line of best fit, what would be the shoe size of a person who is 5.9 feet tall?

    d. According to the line of best fit, what is the height of a person that has 10.7 as shoe size?

    e. Calculate the residuals from the plot above. What do they represent? Are the points with positive residuals located above or below the regression line?

    f. What is the sum of the squared residuals of the linear model that represents the situation described above? Is there any different line that gives a smaller sum? Explain your reasoning.

    g. What is the Least Squares Regression line that models this data? How do you know if this equation is the line of best fit to model the data?

    h. Using technology, compute the correlation coefficient and interpret what it means in the context of the data. (What does the correlation coefficient measure for a linear association?)

    i. Analyze correlation and causation for this data set.

3. The data below shows the approximate number of chess games played and the FIDE rating of 10 chess players.

Number of games	21000	28000	39000	47000	58000	54000	74000	72000	75000	90000
FIDE rating	1545	1635	1710	1850	1925	2025	2300	2380	2410	2750

    a. Represent the data on a scatter plot and describe how the variables are related using the shape, strength and direction of the scatter plot.

    b. Find the linear function that best fits this data. What is the meaning of the slope and the Y-intercept of this linear function in the context of this data?

    c. According to the line of best fit, what would be the FIDE rating of a chess player who played approximately 60,000 chess games in his life?

    d. According to the line of best fit, approximately how many games does a chess player have to play in order to reach a 2,500 FIDE rating?

    e. Calculate the residuals from the plot above. What do they represent? Are the points with positive residuals located above or below the regression line?

f. What is the sum of the squared residuals of the linear model that represents the situation described above? Is there any different line that gives a smaller sum? Explain your reasoning.

g. What is the Least Squares Regression line that models this data? How do you know if this equation is the line of best fit to model the data?

h. Using technology, compute the correlation coefficient and interpret what it means in the context of the data. (What does the correlation coefficient measure for a linear association?)

i. Analyze correlation and causation for this data set.

4. The data below shows the final grades for Algebra 1 and English 1 of 15 students.

Algebra 1	82	96	80	84	65	70	78	81	83	68	60	87	90	94	82
English 1	78	82	83	81	68	71	81	98	92	72	65	84	85	88	93

a. Represent the data on a scatter plot and describe how the variables are related using the shape, strength and direction of the scatter plot.

b. Find the linear function that best fits this data. What is the meaning of the slope and the Y-intercept of this linear function in the context of this data?

c. According to the line of best fit, what would be the final English 1 grade for a student with a final Algebra 1 grade of 85?

d. According to the line of best fit, what would be the final Algebra 1 grade for a student with a final English 1 grade of 75?

e. Calculate the residuals from the plot above. What do they represent? Are the points with positive residuals located above or below the regression line?

f. What is the sum of the squared residuals of the linear model that represents the situation described above? Is there any different line that gives a smaller sum? Explain your reasoning.

g. What is the Least Squares Regression line that models this data? How do you know if this equation is the line of best fit to model the data?

h. Using technology, compute the correlation coefficient and interpret what it means in the context of the data. (What does the correlation coefficient measure for a linear association?)

i. Analyze correlation and causation for this data set.

5. The data below shows the number of hours per day spent navigating on Internet and the number of hours per day spent playing sports by 12 people.

Hours per day for Internet	8	5	10	3	7	6	2	1	5.5	0	9	4
Hours per day for Sports	1	2	1.5	5	6	4.5	5.5	6	2.5	4	3.5	3

a. Represent the data on a scatter plot and describe how the variables are related using the shape, strength and direction of the scatter plot.
b. Find the linear function that best fits this data. What is the meaning of the slope and the Y-intercept of this linear function in the context of this data?
c. According to the line of best fit, how many hours per day would a person spend playing sports if he spends 2 and a half hours per day navigating on Internet?
d. According to the line of best fit, how many hours per day would a person spend navigating on Internet if he spends 7 hours per day playing sports?
e. Calculate the residuals from the plot above. What do they represent? Are the points with positive residuals located above or below the regression line?
f. What is the sum of the squared residuals of the linear model that represents the situation described above? Is there any different line that gives a smaller sum? Explain your reasoning.
g. What is the Least Squares Regression line that models this data? How do you know if this equation is the line of best fit to model the data?
h. Using technology, compute the correlation coefficient and interpret what it means in the context of the data. (What does the correlation coefficient measure for a linear association?)
i. Analyze correlation and causation for this data set.

6. The data below shows the number of hours per day spent navigating on Internet and the annual income (in dollars) of 8 people.

Hours per day for Internet	11	6	3	5	2	8	1	9
Annual income	65000	25000	32000	28000	36000	40000	57000	15000

a. Represent the data on a scatter plot and describe how the variables are related using the shape, strength and direction of the scatter plot.
b. Find the linear function that best fits this data. What is the meaning of the slope and the Y-intercept of this linear function in the context of this data?
c. According to the line of best fit, what would be the annual income of a person that spends 4 hours per day navigating on Internet?
d. According to the line of best fit, how many hours per day does a person spend navigating on Internet if his annual income is $35,000?
e. Calculate the residuals from the plot above. What do they represent? Are the points with positive residuals located above or below the regression line?
f. What is the sum of the squared residuals of the linear model that represents the situation described above? Is there any different line that gives a smaller sum? Explain your reasoning.
g. What is the Least Squares Regression line that models this data? How do you know if this equation is the line of best fit to model the data?

h. Using technology, compute the correlation coefficient and interpret what it means in the context of the data. (What does the correlation coefficient measure for a linear association?)

i. Analyze correlation and causation for this data set.

7. The data below shows the number of hours spent studying for the Algebra 1 final exam and the final exam grade of 10 students.

Number of hours	0	1	2	3	4	5	6	7	8	9
Final Exam grade	25	42	53	57	62	73	75	84	87	93

According to the line of best fit, what is the approximate increase in the final exam grade per hour studied for the exam?

QUIZ #1

N-RN.1 *Explain how the definition of the meaning of rational exponents follows from extending the properties of integer exponents to those values, allowing for a notation for radicals in terms of rational exponents.*

N-RN.2 *Rewrite expressions involving radicals and rational exponents using the properties of exponents.*

A-SSE.2 *Use the structure of an expression to identify ways to rewrite it.*

Simplify: (write the answer in exponential form as well as in root form)

1. $m^{\frac{2}{3}} \cdot m^{\frac{5}{6}}$

2. $\dfrac{a^{\frac{3}{10}}}{a^{\frac{2}{5}}}$

3. $\left(x^{\frac{3}{4}}\right)^{\frac{4}{9}}$

4. $\left(x^{\frac{2}{3}}y^{\frac{3}{5}}z^{4}\right)^{15}$

5. $5\sqrt[4]{6} \cdot 2\sqrt[4]{12} \cdot \sqrt[4]{18}$

6. $9^{\frac{1}{3}} \cdot \sqrt[4]{27}$

7. $\left(x^{2}y^{\frac{3}{5}}z^{4}\right)^{\frac{10}{3}} \cdot \left(x^{\frac{1}{2}}yz^{\frac{2}{3}}\right)^{6}$

8. $(8x)^{\frac{1}{3}}\left(x^{\frac{2}{3}}\right)$

9. $\left(3t^{\frac{1}{3}} \cdot 7n^{\frac{3}{4}}\right)\left(3t^{\frac{5}{6}} \cdot 7n^{\frac{1}{2}}\right)$

QUIZ #2

N-Q.1 Use units as a way to understand problems and to guide the solution of multi-step problems; choose and interpret units consistently in formulas; choose and interpret the scale and the origin in graphs and data displays.

N-Q.2 Define appropriate quantities for the purpose of descriptive modeling.

N-Q.3 Choose a level of accuracy appropriate to limitations on measurement when reporting quantities.

1. An athlete runs 16 miles in 3 hours. Use dimensional analysis to convert the athlete's speed to feet per second.

2. On a map, the distance from New York to Washington DC is 1.75 inches. What is the actual distance, knowing that the scale of the map is 1 inch to 150 miles?

3. Choose a proper scale and graph the following data:

X	0	100	200	300	400	500	600	700
y	0	3	6	9	12	15	18	21

4. Two stores have sales on clothing. A same sweater before sale cost $48. The first store offers a $10 off from the original price and the second store offers 25% off from the original price. Which store offers the better deal?

5. In one year, the price of the gasoline rose from $3.20 per gallon to $3.60 per gallon. What was the percent increase?

6. John's height is measured as 54 inches to the nearest inch. What is John's minimum and maximum possible height?

7. An airplane flies 25 feet in 3 seconds. What is the airplane's speed in miles per hour?

8. The area of a rectangle is at most 40 square inches. The length of the rectangle is 8 inches. What are the possible measurements for the width of the rectangle?

9. Marcos bought a shirt for $45, but the price he paid was 30% off the original price. What was the initial price of the shirt?

QUIZ #3

A-APR.1 Understand that polynomials form a system analogous to the integers, namely, they are closed under the operations of addition, subtraction, and multiplication; add, subtract, and multiply polynomials.

A-SSE.2 Use the structure of an expression to identify ways to rewrite it.

A-SSE.3 Choose and produce an equivalent form of an expression to reveal and explain properties of the quantity represented by the expression.

    a.   *Factor a quadratic expression to reveal the zeros of the function it defines.*

1. If $f = 4x^2 + x + 2$ and $g = 5x^2 + 3x + 1$ are two polynomials, find $3 \cdot f - 2 \cdot g$.
2. The perimeter of a *rectangle* is $(2x^2 - 9x + 7)$ and the length of it is $(x^2 + 2x - 3)$. What is the width of the rectangle in terms of x?
3. Simplify each product:
   - **a.** $-3x \cdot (2x^2 - 7 + 2x)$
   - **b.** $(7 - 4b) \cdot (2b - 5)$
4. Simplify:
   - **a.** $x \cdot (x - 4) - x \cdot (x + 5) + 2x \cdot (x - 3)$
   - **b.** $(3a + 1)(2a - 5) - (a - 3)(2a + 5)$
5. The width of a *rectangle* is $(4x + 1)$ and the length is $(2x + 5)$. What is the area of the rectangle in terms of x?
6. Factor, using the Greatest Common Factor:
   - **a.** $8x^3 + 16x^2 - 24x + 40x^4$
   - **b.** $12x^2y + 8x^3y^3 - 24xy + 16x^2y^2$
7. Factor, using the "*Difference of Two Squares*" formula:
   - **a.** $x^2 - 4$
   - **b.** $x^4 - y^4$
8. Factor the following *Quadratic* expressions:
   - **a.** $x^2 - 5x - 14$
   - **b.** $4x^2 + 12x - 7$
9. Factor by grouping terms:
   - **a.** $x^3 + x^2 - 4x - 4$

QUIZ #4

A-CED.1 Create equations and inequalities in one variable and use them to solve problems. Include equations arising from linear and exponential functions.

A-CED.4 Rearrange formulas to highlight a quantity of interest, using the same reasoning as in solving equations.

A-REI.1 Explain each step in solving a simple equation as following from the equality of numbers asserted at the previous step, starting from the assumption that the original equation has a solution. Construct a viable argument to justify a solution method.

A-REI.3 Solve linear equations and inequalities in one variable, including equations with coefficients represented by letters.

1. Solve each equation:
   a. $3 - 5a = -12$
   b. $-10 = \frac{2}{3}p + 17$

2. Solve each equation:
   a. $2x - 3 - 7x = 10 - 4x + 6$
   b. $10b - 4 \cdot (b + 1) = 2 \cdot (3b + 5) - 5 \cdot (2b + 3) - 26$

3. Solve each equation for the specified variable:
   a. $5ab - 3bc = 10$ for c
   b. $3d = \frac{a-b}{b-c}$ for a

4. If $d = \frac{a-2b}{c}$ is a given equation:
   a. Find the value of a, knowing that $b = -1$, $c = 5$ and $d = -3$.
   b. Find the value of b, knowing that $a = 3$, $c = -2$ and $d = 1$.
   c. Find the value of c, knowing that $a = -6$, $b = 4$ and $d = -5$.
   d. Solve the formula for a.
   e. Solve the formula for b.
   f. Solve the formula for c.

5. If $(3m + 2)x - 4x = 2m(6x - 1) + 8m$ is a given equation:
   a. Find the value of x, knowing that $m = 3$.
   b. Find the value of m, knowing that $x = -5$.
   c. Solve the equation for x.
   d. Solve the equation for m.

6. The sum of three numbers is 1032. Find the numbers, knowing that the second number is three times the first number and the third number is 60 more than half of the first number.

7. The length of a rectangle is $(5x + 8)$ and the width is $(2x + 1)$. What are the possible values for x, if the perimeter of the rectangle is less than 74 yards?

8. Ken earns a base salary of $800 per month as a salesman. In addition to the salary, he earns $75 per product he sells. If his goal is to earn $2300 per month, create an equation to find how many products does he need to sell in order to reach his goal.

QUIZ #5

A-CED.2 Create equations in two or more variables to represent relationships between quantities; graph equations on coordinate axes with labels and scales.

A-REI.10 Understand that the graph of an equation in two variables is the set of all its solutions plotted in the coordinate plane, often forming a curve (which could be a line).

1. Graph the equation $y = 2^x$ and find five solutions to this equation. How many solutions does the equation have? How can the solutions be represented on a graph?

2. Graph the equation $2x - 3y + 6 = 0$ and find five solutions to this equation. How many solutions does the equation have? How can the solutions be represented on a graph?

3. Graph the equation $y = -2 \cdot \left(\frac{1}{2}\right)^x$ and find five solutions to this equation. How many solutions does the equation have? How can the solutions be represented on a graph?

4. Graph the equation $y = 4x + 3$ and find five solutions to this equation. How many solutions does the equation have? How can the solutions be represented on a graph?

5. Graph the equation $x = 2y - 8$ and find five solutions to this equation. How many solutions does the equation have? How can the solutions be represented on a graph?

6. Graph the equation $x = 10 - 4y$ and find five solutions to this equation. How many solutions does the equation have? How can the solutions be represented on a graph?

7. Felix has $200 to buy books and movie DVD's. Knowing that a book costs $8 and a movie DVD costs $20, create an equation to represent the relationship between these quantities. Graph this equation and find the viable solutions to the problem. What is the maximum number of books that Felix can buy? What is the maximum number of movie DVD's that Felix can buy?

8. Tatiana has $3,000 in her savings account and she deposits $200 every month. Create an equation in two variables to represent this situation. Graph this equation and find how much money will Tatiana have in her savings account after 3 years.

9. Elis has $36 to buy sodas and ice-creams. Knowing that a soda costs $2 and an ice-cream costs $3, create an equation to represent the relationship between these quantities. Graph this equation and find the viable solutions to the problem. What is the maximum number of sodas that Elis can buy? What is the maximum number of ice-creams that Elis can buy?

QUIZ #6

A-REI.5 Prove that, given a system of two equations in two variables, replacing one equation by the sum of that equation and a multiple of the other produces a system with the same solutions.

A-REI.6 Solve systems of linear equations exactly and approximately (e.g., with graphs), focusing on pairs of linear equations in two variables.

A-REI.12 Graph the solutions to a linear inequality in two variables as a half- plane (excluding the boundary in the case of a strict inequality), and graph the solution set to a system of linear inequalities in two variables as the intersection of the corresponding half-planes.

A-CED.3 Represent constraints by equations or inequalities, and by systems of equations and/or inequalities, and interpret solutions as viable or non- viable options in a modeling context.

1. Solve the following systems of equations:

    a. $\begin{cases} 6x - 2y = -14 \\ 7x + y = -13 \end{cases}$

    b. $\begin{cases} 2(x + 1) - 3(y - 1) = -1 \\ 5(x - 2) - 4(y - 3) = 1 \end{cases}$

2. Solve the following systems of equations using *graphs* (graph both linear equations and then find the coordinates of the intersection point of the two graphs):

    a. $\begin{cases} y = 2x - 6 \\ y = -3x + 4 \end{cases}$

    b. $\begin{cases} 4x - 3y = 20 \\ 5x + 2y = 2 \end{cases}$

3. Graph the solution to the following systems of linear inequalities:

    a. $\begin{cases} y \le 2x - 6 \\ y \ge -3x + 4 \end{cases}$

    b. $\begin{cases} x + 3y - 6 \ge 0 \\ x \le 4 \\ y > -2 \end{cases}$

4. Find two numbers, knowing that their sum is 1240 and one of the numbers is 24 more than three times the other number.

5. In a classroom there are 32 students, boys and girls. If other 5 boys would come in the classroom and 7 girls would leave, then the number of boys would double the number of girls.

    a. Write a system of equations for the context of this problem.

    b. How many boys and girls are there in the classroom?

    c. Is the solution of the system viable for this situation? Explain your reasoning.

6. The sum of two numbers is 623 and the difference of them is 73. Find the numbers.

7. Emilia has in her wallet three times as many 5-dollar bills as 10-dollar bills. She has a total of $200. How many 10-dollar bills does Emilia have in her wallet?

8. Felix wants to buy books and notebooks from a store. A book costs $8 and a notebook costs $3. Felix has $48 to spend.
   a. Construct a linear inequality for this situation.
   b. What are the possible combinations of books and notebooks that he can buy? Explain your reasoning.
   c. Graph the linear inequality that represents this situation. Interpret the solution and find the viable solutions in this context.

9. Felix needs to earn at least $850 a week to be able to pay his mortgage and his bills, but he cannot work more than 60 hours per week. Felix is getting paid $20 per hour for mowing lawns and $10 per hour for washing cars.
   a. Create a system of inequalities that represents this situation.
   b. Graph the system and interpret the solution. Find the viable solution in this context.

QUIZ #7

A-REI.11 Explain why the x-coordinates of the points where the graphs of the equations y = f(x) and y = g(x) intersect are the solutions of the equation f(x) = g(x); find the solutions approximately, e.g., using technology to graph the functions, make tables of values, or find successive approximations. Include cases where f(x) and/or g(x) are linear and exponential functions.

1. Solve the equation $3^x = -x + 4$ using graphs:
   a. Graph the equations $y = 3^x$ and $y = -x + 4$ in the same coordinate plane.
   b. In how many points do these graphs intersect? What do these points of intersection represent?
   c. Using a table of values for both equations, find approximately the X-coordinate of the point where the graphs intersect.
   d. Using technology, find exactly the coordinates of the point where the graphs of the equations intersect.
   e. What is the solution of the equation $3^x = -x + 4$? What is the solution of the system of equations: $\begin{cases} y = 3^x \\ y = -x + 4 \end{cases}$? Explain why the X-coordinates of the points where the graphs of the equations $y = 3^x$ and $y = -x + 4$ intersect are the solutions of the equation $3^x = -x + 4$.

2. Solve the equation $5 \cdot 2^x = 4 \cdot \left(\frac{1}{3}\right)^x$ using graphs:
   a. Graph the equations $y = 5 \cdot 2^x$ and $y = 4 \cdot \left(\frac{1}{3}\right)^x$ in the same coordinate plane.
   b. In how many points do these graphs intersect? What do these points of intersection represent?
   c. Using a table of values for both equations, find approximately the X-coordinate of the point where the graphs intersect.
   d. Using technology, find exactly the coordinates of the point where the graphs of the equations intersect.
   e. What is the solution of the equation $5 \cdot 2^x = 4 \cdot \left(\frac{1}{3}\right)^x$? What is the solution of the system of equations: $\begin{cases} y = 5 \cdot 2^x \\ y = 4 \cdot \left(\frac{1}{3}\right)^x \end{cases}$? Explain why the X-coordinates of the points where the graphs of the equations $y = 5 \cdot 2^x$ and $y = 4 \cdot \left(\frac{1}{3}\right)^x$ intersect are the solutions of the equation $5 \cdot 2^x = 4 \cdot \left(\frac{1}{3}\right)^x$.

QUIZ #8

F-IF.1 Understand that a function from one set (called the domain) to another set (called the range) assigns to each element of the domain exactly one element of the range. If f is a function and x is an element of its domain, then f(x) denotes the output of f corresponding to the input x. The graph of f is the graph of the equation y = f(x).

F-IF.2 Use function notation, evaluate functions for inputs in their domains, and interpret statements that use function notation in terms of a context.

1. If $f: A \to B$ is a function, and the domain $A = \{1, 2, 3, 4\}$ and $f(x) = x^2 + 2x - 3$, find set B (the range). Complete a table and graph the function f.

2. If $f: A \to B$ is a function, and the range $B = \{1, 2, 3, 4\}$ and $f(x) = 2x - 5$, find set A (the domain). Complete a table and graph the function f.

3. Identify the independent and dependent variables. Write an equation in function notation for each situation:
   a. A lawyer's fee is $150 per hour for her services;
   b. The admission fee to an amusement park is $10. Each ride costs $3.

4. A swimming pool containing 20,000 gallons of water is being drained. Every hour, the volume of the water in the pool decreases by 750 gallons.
   a. Write an equation to describe the volume V of water in the pool after h hours.
   b. How much water is in the pool after 2 hours?
   c. Create a table of values showing the volume of the water in gallons in the swimming pool as a function of the time in hours and graph the function.

5. If $f(x) = 2x - 5$ is a function, solve for x:
$$3 \cdot f(x) - 2 = 7$$

6. The function $p(t) = 2.7 \cdot (1.08)^t$ gives a city's population $p(t)$ (in millions), where t is the number of years since 1998. According to this function, what was the population of the city in 2010?

7. If $f(x) = 2x - 5$ is a function, evaluate $f(4)$.

8. Evaluate $h(x) = \frac{3}{4} \cdot (5 - 8x) + 3x$ when $x = \frac{3}{8}$.

9. If $f(x) = \begin{cases} 4x - 3, & \text{if } x \le 2 \\ -2x + 5, & \text{if } x > 2 \end{cases}$ is a function, evaluate $f(-3), f(4)$.

QUIZ #9

F-IF.3 Recognize that sequences are functions, sometimes defined recursively, whose domain is a subset of the integers.

1. Give an example of a sequence defined recursively. Explain the meaning of the word "recursive".

2. Give an example of a sequence defined explicitly.

3. If $a_n = 2n + 5$ is a sequence, find a_3, a_4, a_{10}.

4. If $x_n = 3 \cdot x_{n-1} + 5$ is a sequence defined recursively and $x_1 = 9$, find x_2, x_5, x_6.

5. Write a recursive formula for each sequence:
   a. 19, 15, 11, 7,...
   b. 2, 6, 18, 54, 162,...

6. For each recursive formula, write an explicit formula:
   a. $a_1 = 7, a_n = a_{n-1} - 3$
   b. $a_1 = 64, a_n = \frac{1}{4} \cdot a_{n-1}$

7. For each explicit formula, write a recursive formula:
   a. $a_n = 2 \cdot 3^{n-1}$
   b. $a_n = 7n + 1$

8. The balance B_{n+1} in Mr. Ackerman's savings account at the end of a year is calculated by the equation $B_{n+1} = 1.028 \cdot B_n$, where B_n is the balance at the end of the previous year. Mr. Ackerman made a deposit to open the account 4 years ago. He has not made any additional deposits or withdrawals since. The balance at the end of 2 years was $9,537.14. What is the balance at the end of 4 years?

9. The number of bacteria in a culture can be represented by the formula $N_t = 3.7 \cdot N_{t-1}$. In the formula, N_t is the number of bacteria at the end of t minutes, and N_{t-1} is the number of bacteria at the end of $t - 1$ minutes. There are 8,417 bacteria in the culture at the end of 5 minutes. How many bacteria will be in the culture at the end of 9 minutes?

QUIZ #10

F-IF.4 *For a function that models a relationship between two quantities, interpret key features of graphs and tables in terms of the quantities, and sketch graphs showing key features given a verbal description of the relationship. Key features include: intercepts; intervals where the function is increasing, decreasing, positive, or negative; relative maximums and minimums; symmetries.*

F-IF.5 *Relate the domain of a function to its graph and, where applicable, to the quantitative relationship it describes.*

1. Find the X and Y intercepts of the following functions:
    a. $f(x) = 2x - 4$
    b. $f(x) = x^2 + 2x - 3$
    c. $f(x) = 3 \cdot e^{2x}$

2. Given the graph below:

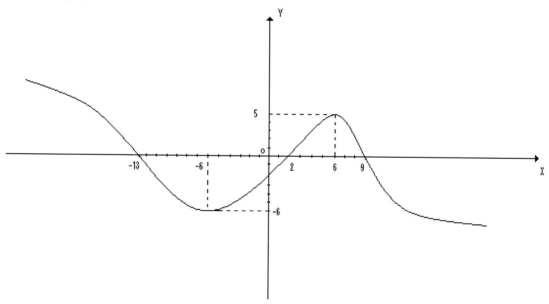

    a. What are the X intercepts of the function f? What is the Y intercept?
    b. Can a function have more than one Y intercept? Explain your answer.
    c. Where is the function f increasing? Decreasing?
    d. Where is the function f positive? Negative?
    e. Specify the relative minimums and maximums of the function f.
    f. Is there any symmetry? Explain your answer.

3. Find all the key features (X intercepts, Y intercept, relative minimums, relative maximums, where is the function increasing/decreasing, positive or negative) for the following functions (using maybe the graphing calculator):
   a. $f(x) = 3x + 6$
   b. $f(x) = 2 \cdot e^{4x}$
   c. $f(x) = x^2 + 3x - 4$

4. The gas tank in Felix's truck holds 25 gallons, and the truck can travel 15 miles for each gallon of gas. When Felix begins with a full tank of gas, the function $f(x) = -\frac{1}{15}x + 25$ gives the amount of gas $f(x)$ that will be left in the tank after traveling x miles (if he does not buy more gas).
   a. What is the practical domain and range for this function?
   b. Graph the function on the practical domain.

5. An arrow is launched from 2 meters above the ground at time $t = 0$. The function that models this situation is given by $h(t) = -\frac{1}{16}t^2 + \frac{7}{8}t + 2$, where t is the time measured in seconds and h is the height above the ground measured in meters.
   a. What is the practical domain for t in this context?
   b. What is the height of the arrow three seconds after it was launched?
   c. What is the maximum value of the function and what does it mean in context?
   d. When is the arrow 3 meters above the ground?
   e. When is the arrow 1 meter above the ground?
   f. What are the intercepts of this function? What do they mean in the context of this problem?
   g. What are the intervals of increase and decrease on the practical domain? What do they mean in the context of this problem?

QUIZ #11

F-IF.6 Calculate and interpret the average rate of change of a function (presented symbolically or as a table) over a specified interval. Estimate the rate of change from a graph.

F-IF.7 Graph functions expressed symbolically and show key features of the graph, by hand in simple cases and using technology for more complicated cases.

    a. Graph linear and quadratic functions and show intercepts, maxima, and minima.
    e. Graph exponential functions, showing intercepts and end behavior.

1. The table shows the balance of a bank account on different days of the month.

Day	1	5	10	15	22	30
Balance	2500	2000	1730	1300	1000	720

    a. Find the rate of change for each time interval.
    b. During which time interval did the balance decrease at the greatest rate?

2. Use rates of change to determine whether the following functions are linear or non-linear. (*If the function is linear then the rate of change of the function is the same as the slope of the line that represents the graph of the function*)

x	-2	1	4	8	12
y	5	3	-2	10	15

3. Find the average rate of change of the following function over the specified interval:
$$f(x) = -3x + 5 \text{ ; interval } [1,7]$$

4. Find the average rate of change of the following function over the specified interval:
$$f(x) = 3 \cdot 2^x \text{ ; interval } [0,2]$$

5. Find the average rate of change of the following function over the specified interval:
$$f(x) = x^2 - 4x + 5 \text{ ; interval } [-4,9]$$

6. Felix has $5000 to invest in one of the two financial plans offered by a bank. Plan A offers to increase his principal by $180 each year, while plan B offers to pay 3.8% interest compounded quarterly. The amount of each investment after t years, is given by: $A = 5000 + 180t$ and $B = 5000 \cdot (1.0095)^{4t}$ respectively.

   What plan should Felix choose if he invests his money for 5 years? What if he invests his money for 10 years? Explain your reasoning, using function values, the average rate of change and the graphs of the equations.

7. Graph the following function and find the X and Y intercepts of the graph:

$$f(x) = -\frac{2}{5}x + 10$$

8. For the following Quadratic Function find the Vertex, the X intercepts, the Y intercept and graph the function. After graphing, find the axis of symmetry, the minimum or maximum, the domain and the range of the function:

$$y = -3x^2 + 12x - 5$$

9. Graph the following exponential function, showing intercepts and end behavior:

$$y = -3 \cdot 4^x$$

QUIZ #12

F-IF.8 Write a function defined by an expression in different but equivalent forms to reveal and explain different properties of the function.

    a.   Use the process of factoring a quadratic function to show zeros and interpret these in terms of a context.
    b.   Use the properties of exponents to interpret expressions for exponential functions.

A-SSE.3 Choose and produce an equivalent form of an expression to reveal and explain properties of the quantity represented by the expression.

    c.   Factor a quadratic expression to reveal the zeros of the function it defines.

1. A college's tuition has been increasing 3% each year since 1980. If the tuition cost in 1980 was $7,000, write a function for the amount of the tuition x years after 1980. Find the cost of tuition for this college in 2014.

2. Felix deposited $3,000 in 2009 into a savings amount at a bank that offers 6% per year compounded monthly. How much money will Felix have in 2014?

3. Emilia deposited $7,800 in 2006 into a savings account at a bank that offers 4% per year compounded quarterly. How much money will Emilia have in 2015?

4. What are the zeros of the following function:
$$f(x) = x^2 + 2x - 3$$

5. Find the Quadratic function that has the zeros 1 and -4.

6. The annual tuition at a private school since 2004 is modeled by the function $f(x) = 800 \cdot (1.09)^x$, where x is the number of years since 2004.
   a. What was the tuition cost in 2004?
   b. What is the annual percentage of tuition increase?

7. A company earns a weekly profit of P dollars by selling x items according to the function:
$P(x) = -x^2 + 25x - 100$
   a. Find the zeros of the function P and interpret these zeros in the context of the problem.
   b. How many items does the company have to sell each week to maximize the profit?

8. Suppose $h(t) = -3t^2 + 5t + 4$ is a function giving the height of a diver above the water (in meters), t seconds after the diver leaves the springboard.
   a. How high above the water is the springboard? Explain your reasoning.
   b. When does the diver hit the water?
   c. At what time on the diver's descent toward the water is the diver again at the same height as the springboard?
   d. When does the diver reach the peak of the dive?

9. A certain substance has a half-life of 20 seconds. Find the amount of the substance left from an 80 gram sample after 2 minutes.

QUIZ #13

F-IF.9 Compare properties of two functions each represented in a different way (algebraically, graphically, numerically in tables, or by verbal descriptions).

1. Two phone companies offer the following plans: Company A charges $40 a month and 6 cents per minute of usage, while company B charges according to the linear function represented by the table:

m (minutes)	0	100	200	300	400	500	600
C (cost)	$35	$43	$51	$59	$67	$75	$83

   a. Describe and compare the two plans given above.

   b. If John enjoys talking on the phone, which plan should he choose?

2. Nova Taxi cab company charges a $3 boarding rate in addition to its meter which is $2 for every mile. PRITAX cab company charges according to the function $P = 1.5x + 6$, where x is the number of miles traveled.

   a. Compare the charging policies of the two cab companies.

   b. Which cab company has a better price? Explain your answer.

3. The height, h, in feet of a football above the ground is given by the function $h(t) = -4t^2 + 32t + 3$, where t is the time in seconds. The height of a soccer ball above the ground is represented in the graph below:

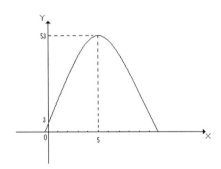

   a. Compare the properties of the two functions given above.

   b. What ball reaches a higher height and which ball stays more in the air?

4. Felix's company weekly revenue in dollars is given by the function $R(x) = 2000x - 2x^2$ where x is the number of items produced during a week. Emilia's company weekly revenue is represented in the graph below:

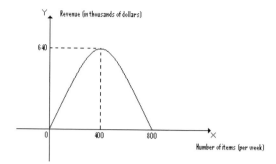

a. What amount of items will produce the maximum revenue in Felix's company? What amount of items will produce the maximum revenue in Emilia's company?

b. Compare the evolution of the revenue of the two companies represented by the functions above.

5. The annual tuition at a college A since 2005 is modeled by the function $T(x) = 3000 \cdot (1.03)^x$, where x is the number of years since 2005. The annual tuition at another college B is represented by a graph below:

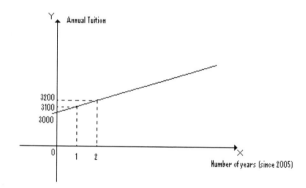

Compare the properties of the two functions and describe the evolution of the tuition cost for the two colleges.

6. Felix wants to invest $5,000 into a financial plan. Three banks offer the following choices as financial investments:

Bank A: $y = 5000 + 200 \cdot x$

Bank B: $y = 5000 + 50x^2 - 200x$

Bank C: $y = 5000 \cdot (1.02)^x$, where y represents the amount of money after x years .

a. Which plan is better for Felix if he invests his money for 1 year? How about 5 years? How about 10 years?

b. Compare the three financial plans using tables or graphs and find which financial plan is the best for different time periods.

QUIZ #14

F-BF.1 Write a function that describes a relationship between two quantities.

    a.   *Determine an explicit expression, a recursive process, or steps for calculation from a context.*

    b.   *Combine standard function types using arithmetic operations.*

1. Felix knows that money in an account where interest is compounded semi-annually will earn interest faster than money in an account where interest is compounded annually. He wonders how much interest can be earned by compounding it more and more often. Let's suppose Felix invests $1 at a 100% interest rate.

    a. What will the year-end balance be if the interest is compounded annually?

    b. What will the year-end balance be if the interest is compounded semi-annually?

    c. What will the year-end balance be if the interest is compounded quarterly?

    d. What will the year-end balance be if the interest is compounded monthly?

    Explain how you calculate the year-end balance in each situation.

2. Using the graph below, sketch a graph of the function $s(x) = f(x) + g(x)$ on interval $[-1,8]$

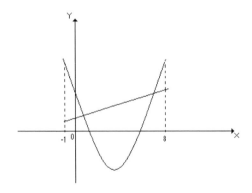

3. The price of the gas in 1980 was $1.50 per gallon and in 2006 the price of the gas was $2.28 per gallon.

    a. Assuming that the increase is linear, write a function that describes the relationship between the price per gallon and the number of years since 1980.

    b. What would be the price of the gas in 2013?

4. A car rental company charges $19.95 per day plus $0.15 per mile.

    a. Write a function that gives the cost in dollars to travel x miles over a 7-day period.

    b. Calculate the cost in dollars to travel 250 miles over a 2-day period.

5. The first term of a sequence is 3, the second term of the sequence is 7, the third term is 11 and so forth.

    a. Determine a recursive process of the sequence.

    b. Write a function that describes the function.

    c. How does the recursive formula relate to the function that describes the sequence?

6. If $f(x) = 2x - 2$ and $g(x) = x^2 + 2x - 3$ are two functions:

    a. Graph the functions f and g.

    b. Find the expression of the function $h(x) = f(x) + g(x)$.

    c. Sketch the graph of the function h using the graphs of the functions f and g.

    d. Graph the function h, using the explicit form found at point (b) and then compare this graph with the graph from point (c).

7. If $f(x) = x^2 + 4x - 5$ and $g(x) = 8$ are two functions:

    a. Graph the functions f and g.

    b. Find the expression of the function $h(x) = f(x) + g(x)$.

    c. Sketch the graph of the function h using the graphs of the functions f and g.

    d. Graph the function h, using the explicit form found at point (b) and then compare this graph with the graph from point (c).

QUIZ #15

F-BF.2 Write arithmetic and geometric sequences both recursively and with an explicit formula, use them to model situations, and translate between the two forms.

F-BF.3 Identify the effect on the graph of replacing f(x) by f(x) + k and f(x + k) for specific values of k (both positive and negative); find the value of k given the graphs. Experiment with cases and illustrate an explanation of the effects on the graph using technology.

1. Write a recursive formula the sequence below (identifying first if it's an arithmetic or geometric sequence)
$$18, 15, 12, 9,...$$

2. Write a recursive formula the sequence below (identifying first if it's an arithmetic or geometric sequence)
$$1, 3, 9, 27, 81,...$$

3. For the explicit formula below, write a recursive formula:
$$a_n = 5 \cdot 2^{n-1}$$

4. For the explicit formula below, write a recursive formula:
$$a_n = 5n + 9$$

5. For the recursive formula below, write an explicit formula:
$$a_1 = 19, a_n = a_{n-1} - 4$$

6. For the recursive formula below, write an explicit formula:
$$a_1 = 200, a_n = 0.2 \cdot a_{n-1}$$

7. Given the graph of $f(x)$ where does the graph of $f(x + 2)$ translate?
8. Given the graph of $f(x)$ where does the graph of $f(x) - 5$ translate?
9. Which equation represents the graph of $f(x) = 2^{x+1}$ translated 4 units to the left and 3 units up?

QUIZ #16

F-LE.1 Distinguish between situations that can be modeled with linear functions and with exponential functions

    a. Prove that linear functions grow by equal differences over equal intervals, and that exponential functions grow by equal factors over equal intervals.

    b. Recognize situations in which one quantity changes at a constant rate per unit interval relative to another.

    c. Recognize situations in which a quantity grows or decays by a constant percent rate per unit interval relative to another.

1. Emilia has $3,000 in her savings account and she deposits $250 every month.
   a. Determine whether this situation represents a linear model or an exponential model. Explain your reasoning.
   b. Find the function rule that models this situation.
   c. Using the function rule, find how much money will Emilia have in her savings account after 2 years.

2. The price of a new car depreciates 12% each year after it is purchased.
   a. Determine whether this situation represents a linear model or an exponential model. Explain your reasoning.
   b. Find the function rule that models this situation.
   c. Using the function rule, find how much does a car cost after 5 years if the car was bought as new for $25,000.

3. Johnny has to repair his car. The shop charges $25 as an initial fee and $15 per hour of labor.
   a. Determine whether this situation represents a linear model or an exponential model. Explain your reasoning.
   b. Find the function rule that models this situation.
   c. Using the function rule, find how much did Johnny pay if his car was fixed in 6 hours.

4. The population of a city is increasing 8% every five years.
   a. Determine whether this situation represents a linear model or an exponential model. Explain your reasoning.
   b. Find the function rule that models this situation.
   c. What will the population of the city be in 2015 if the population in 1950 was 20,000 people?

5. Felix has $10,000 to invest in one of the two financial plans offered by a bank. Plan A offers to increase his principal by $500 each year, while plan B offers to pay 2.5% interest compounded annually.
   a. What plan should Felix choose if he invests his money for 5 years? What if he invests his money for 10 years?

b. Write a function rule that gives the balance after x years for plan A as well as for plan B.

c. Compare the two functions, determine whether they represent a linear or an exponential model and find which plan is a better deal for Felix and for what period of time.

6. Construct an exponential function knowing that the graph of this function contains the points $A(3, 4)$ and $B(5, 16)$.

7. Construct a linear function knowing that the graph of this function contains the points $A(-2, 5)$ and $B(-7, -1)$.

8. Construct a linear function that has the graph below:

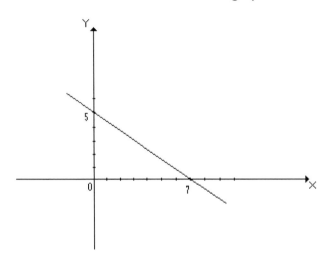

9. Construct an exponential function that has the graph below:

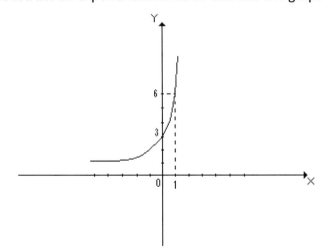

QUIZ #17

F-LE.3 Observe using graphs and tables that a quantity increasing exponentially eventually exceeds a quantity increasing linearly or quadratically.

1. Bobby wants to invest $5,000 into a financial plan. Three banks offer the following choices as financial investments:

   Bank A: $y = 5000 + 200 \cdot x$

   Bank B: $y = 5000 + 50x^2 - 200x$

   Bank C: $y = 5000 \cdot (1.02)^x$, where y represents the amount of money after x years .

   a. Which plan is better for Bobby if he invests his money for 1 year? How about 5 years? How about 10 years?

   b. Compare the three financial plans using tables or graphs and find which financial plan is the best for different time periods.

2. If $f(x) = 8x + 3$, $g(x) = x^2 - 2x + 3$ and $h(x) = 3 \cdot (1.36)^x$ are three functions:

   a. What is the Y intercept of each function? What do you observe about the Y intercepts?

   b. Graph the functions f, g and h in the same system of coordinates.

   c. Find the interval where function f is greater than the functions g and h.

   d. Find the interval where function g is greater than the functions f and h.

   e. Find the interval where function h is greater than the functions f and g.

   f. What do you observe about the function h? Is it true that the exponential function exceeds the linear and the quadratic function after a certain value?

QUIZ #18

F-LE.5 *Interpret the parameters in a linear or exponential function in terms of a context.*

A-SSE.1 *Interpret expressions that represent a quantity in terms of its context.*

   a. Interpret parts of an expression, such as terms, factors, and coefficients.
   b. Interpret complicated expressions by viewing one or more of their parts as a single entity.

1. For the equation $y = m \cdot x + b$ what is the meaning of parameter m and what is the meaning of parameter b?

2. For the equation $y = a \cdot b^x$ what is the meaning of parameter a and what is the meaning of parameter b?

3. The equation $y = 350x + 4700$ represents the value (in dollars) of a work of art from 1958 to 2006. What does the number 350 represent in this context? How about 4700?

4. The function $f(x) = -\frac{1}{15}x + 20$ gives the amount of gas $f(x)$ that will be left in the tank of a truck after traveling x miles. Interpret the coefficients $-\frac{1}{15}$ and 20 in the context of this problem.

5. Juan is going to an amusement park. The equation $A = 300 - 12R$ represents the amount of money, A, he has left after R rides. Interpret the coefficients 300 and -12 in the context of this problem.

6. The annual tuition at a private school since 2001 is modeled by the function $f(x) = 7000 \cdot (1.06)^x$, where x is the number of years since 2004. Interpret the coefficient 7,000 and the base of the exponent 1.06 in the context of this problem.

7. Write the algebraic expression for each verbal expression:
    a. Two times the sum of x and y
    b. The difference between one third of a number and 10
    c. The opposite of the reciprocal of a number

8. The value of Mr. Johnson's car, x years after it is purchased is given by the function $V(x) = 14,000(0.91)^x$. What does the coefficient 14,000 represent in the context of this problem? What does the factor 0.91 represent in the context of this problem?

9. Write a verbal expression for each algebraic expression:
    a. $4 \cdot (a + 9)$
    b. $5x + 3$

QUIZ #19

G-GPE.4 Use coordinates to prove simple geometric theorems algebraically. For example, prove or disprove that a figure defined by four given points in the coordinate plane is a rectangle; prove or disprove that the point $(1, \sqrt{3})$ lies on the circle centered at the origin and containing the point $(0, 2)$.

G-GPE.5 Prove the slope criteria for parallel and perpendicular lines and use them to solve geometric problems (for example, find the equation of a line parallel or perpendicular to a given line that passes through a given point).

1. If $x + 5y - 10 = 0$ is the equation of a line:
   a. Find the coordinates of two different points on this line and plot these points.
   b. Using the coordinates of the two points found at point (a), find the slope of this line.
2. If $y = 7x - 13$ is the equation of a line:
   a. Find the coordinates of two different points on this line and plot these points.
   b. Using the coordinates of the two points found at point (a), find the slope of this line.
3. Find the distance between the two given points:
   $$G(-3, -7) \quad \text{and} \quad H(5, -6)$$
4. Find the slope of the line PQ, knowing the coordinates of P and Q: $P(-2,7)$ and $Q(8, -4)$.
5. Find the slope of the lines given by the equation:
   a. $y = 4x + 7$
   b. $3x - 5y = 14$
6. Using the *slope criteria*, tell whether each pair of lines is *parallel, perpendicular* or *neither*:
   a. $y = 3x + 5$ and $y = 3x - 10$
   b. $y = -\frac{1}{3}x + 8$ and $2x + 6y = 17$
7. Write an equation for the line that is *parallel* to the given line and that passes through the given point:
   $$y = 4x + 5, \qquad A(3,7)$$
8. Write an equation for the line that is *perpendicular* to the given line and that passes through the given point:
   $$2x - 3y = 9, \qquad C(4, -2)$$
9. Find the value of parameter m knowing that the line $y = 4x - 3$ is *perpendicular* to the line $y = (m + 1)x - 2$.

QUIZ #20

G-GPE.6 Find the point on a directed line segment between two given points that partitions the segment in a given ratio.

G-GPE.7 Use coordinates to compute perimeters of polygons and areas of triangles and rectangles, for example, using the distance formula.

G-GMD.1 Give an informal argument for the formulas for the circumference of a circle, area of a circle, volume of a cylinder, pyramid, and cone. Use dissection arguments, Cavalieri's principle, and informal limit arguments.

G-GMD.3 Use volume formulas for cylinders, pyramids, cones, and spheres to solve problems.

1. Find the coordinates of the midpoint of segment CD, if $C(4, -6)$ and $D(-10,2)$.
2. If $M(-3,4)$ is the midpoint of segment PQ and $P(7, -1)$, what are the coordinates of Q?
3. On a map, Sofia's house is located at $(-9,1)$ and Tatiana's house is located at $(5, -8)$. What are the coordinates for Diana's house if she lives exactly halfway between Sofia and Tatiana? What is the distance between Sofia's house and Diana's house?
4. Find the perimeter of the triangle ΔMNP with the vertices $M(2,6)$, $N(7, -1)$ and $P(-5, -3)$.
5. Find the area of the rectangle $ABCD$ with the vertices $A(3,1)$, $B(5, -3)$, $C(9, -1)$ and $D(7,3)$.
6. Find the area of the triangle ΔABC with the vertices $A(-8, 5)$, $B(7, -2)$ and $C(5,10)$.
7. Mr. Johnson wants to build a square pyramid out of cement. He wants the pyramid to be 8 feet tall and its base edge to be 12 feet long. How many cubic feet of cement will he need?
8. In order to make an iron ball, a cylindrical iron tube is heated and pressed into a spherical shape with the same volume. The height of the cylindrical tube is 3 inches and the radius is 1 inch. Find the radius of the iron ball.
9. The diameter of Planet Earth is 12756.2 kilometers (7926.3 miles).
   a. What is the surface area of Planet Earth?
   b. What is the volume of Planet Earth?

QUIZ #21

S-ID.1 Represent data with plots on the real number line (dot plots, histograms, and box plots).

S-ID.2 Use statistics appropriate to the shape of the data distribution to compare center (median, mean) and spread (interquartile range, standard deviation) of two or more different data sets.

S-ID.3 Interpret differences in shape, center, and spread in the context of the data sets, accounting for possible effects of extreme data points (outliers).

1. Create a *histogram*, a *box-plot* and a *stem-plot* for the following data that represents the Algebra 1 test scores of 35 students:
   28, 37, 63, 66, 67, 68, 68, 69, 72, 75, 77, 81, 82, 86, 88, 88, 88, 89, 89, 89, 90, 90, 90, 91, 93, 93, 94, 95, 95, 96, 97, 100, 100, 100, 100
   Describe the shape, center and spread of this data.

2. What are the elements of a *box-plot*?

3. The data below shows the number of people living in a house for 15 residences in a small neighborhood of a city:
   2, 1, 5, 1, 3, 4, 3, 2, 4, 2, 4, 2, 3, 4, 3
   Create a *dot plot* to display the given data and then describe the shape, center and spread of this data.

4. The average monthly income in millions of dollars at a Shopping Mall is shown in the table below:

Jan	Feb	Mar	Apr	May	Jun	Jul	Aug	Sep	Oct	Nov	Dec
2.3	1.9	2.1	1.8	2.2	2.2	2.0	2.8	2.2	2.1	2.5	7.9

   a. Represent this data using a histogram and a box-plot.
   b. According to the shape of this distribution, what measure of spread is more appropriate for this data (*Standard Deviation* or *Interquartile Range*)?
   c. According to the shape of this distribution, what measure of center is more appropriate for this data (*Mean* or *Median*)?
   d. Are there any *outliers*? Explain your reasoning.

5. You scored an 87 on a test in your Algebra 1 class where the mean was 85 and the standard deviation was 3. Your best friend is in a different Algebra 1 class and scored a 90 where the mean in her class was 88 and the standard deviation was 4. Who had the better score relative to their own class?

6. The data below shows the customer ratings for 2 hotel resorts.
   Hotel A: 4.3, 4.1, 3.7, 4.5, 3.2, 4.2, 4.3, 3.8, 4.0, 3.1, 4.9, 3.8, 3.2, 3.6, 4.1
   Hotel B: 4.4, 2.5, 3.9, 3.7, 3.8, 2.9, 4.3, 3.6, 4.1, 2.8, 4.9, 3.4, 3.8, 4.2, 3.9
   a. Construct parallel box-plots of the two distributions.
   b. Create a histogram for each distribution.
   c. Compare center and spread of the two data sets using statistics appropriate to the shape of the distributions.

7. Compare center and spread of the two distributions given:

a.

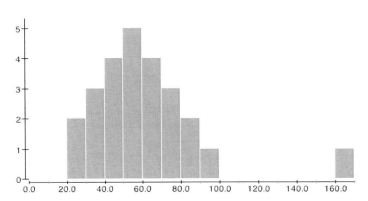

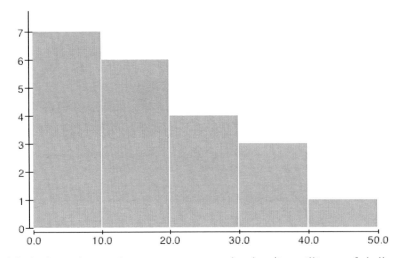

8. The table below shows the average annual sales (in millions of dollars) at a Shopping Center for the last 10 years.

Year	2002	2003	2004	2005	2006	2007	2008	2009	2010	2011
Sales (in millions)	9.3	9.8	8.7	15.9	7.2	6.9	5.8	7.5	5.4	7.0

    a. Find the mean and the standard deviation of the data.
    b. In what years were the sales less than one standard deviation from the mean?
    c. Find the median and interquartile range for the data.

QUIZ #22

S-ID.5 Summarize categorical data for two categories in two-way frequency tables. Interpret relative frequencies in the context of the data (including joint, marginal, and conditional relative frequencies). Recognize possible associations and trends in the data.

1. A survey of 20 adults asked: "Of SEDAN, SUV and TRUCK, which is your favorite type of car?" Each respondent's gender was also recorded. The results were as follows:

Type of car	Gender
SEDAN	FEMALE
SUV	FEMALE
TRUCK	MALE
SUV	FEMALE
SUV	MALE
SEDAN	FEMALE
TRUCK	MALE
SUV	FEMALE
TRUCK	FEMALE
SUV	FEMALE
SUV	MALE
TRUCK	MALE
SUV	FEMALE
SEDAN	FEMALE
TRUCK	MALE
SUV	FEMALE
TRUCK	MALE
SUV	FEMALE
SEDAN	MALE
SUV	FEMALE

a. Create a *two-way frequency table* that summarizes these data including *marginal frequencies*. Interpret the *joint frequencies* of this table.

b. Create a *two-way relative frequency table* that summarizes these data which displays *conditional frequencies* for **gender**. Interpret the *conditional frequencies* of this table.

c. Create a *two-way relative frequency table* that summarizes these data which displays *conditional frequencies* for **type of car**. Interpret the *conditional frequencies* of this table.

2. A survey of 18 people asked: "Of SPRING, SUMMER, FALL and WINTER, which is your favorite season of the year?" Each person's occupation was also recorded. The results were as follows:

Season	Occupation
SPRING	ARCHITECT
SUMMER	TEACHER
WINTER	PHARMACIST
SPRING	ARCHITECT
FALL	PHARMACIST
SUMMER	TEACHER
SUMMER	TEACHER
SPRING	ARCHITECT
WINTER	TEACHER
SUMMER	ARCHITECT
FALL	PHARMACIST
SPRING	PHARMACIST
FALL	PHARMACIST
SUMMER	TEACHER
WINTER	ARCHITECT
FALL	PHARMACIST
SUMMER	TEACHER
FALL	ARCHITECT

a. Create a *two-way frequency table* that summarizes these data including *marginal frequencies*. Interpret the *joint frequencies* of this table.

b. Create a *two-way relative frequency table* that summarizes these data which displays *conditional frequencies* for **occupation**. Interpret the *conditional frequencies* of this table.

c. Create a *two-way relative frequency table* that summarizes these data which displays *conditional frequencies* for **season**. Interpret the *conditional frequencies* of this table.

QUIZ #23

S-ID.6 Represent data on two quantitative variables on a scatter plot, and describe how the variables are related.

   a. Fit a function to the data; use functions fitted to data to solve problems in the context of the data. Use given functions or choose a function suggested by the context. Emphasize linear and exponential models.
   b. Informally assess the fit of a function by plotting and analyzing residuals.
      ***Note:** At this level, for part b, focus on linear models.*

   c. Fit a linear function for a scatter plot that suggests a linear association.
S-ID.7 Interpret the slope (rate of change) and the intercept (constant term) of a linear model in the context of the data.

S-ID.8 Compute (using technology) and interpret the correlation coefficient of a linear fit.

S-ID.9 Distinguish between correlation and causation.

1. The data below shows the final grades for Algebra 1 and English 1 of 15 students.

Algebra 1	82	96	80	84	65	70	78	81	83	68	60	87	90	94	82
English 1	78	82	83	81	68	71	81	98	92	72	65	84	85	88	93

   a. Represent the data on a scatter plot and describe how the variables are related using the shape, strength and direction of the scatter plot.
   b. Find the linear function that best fits this data. What is the meaning of the slope and the Y-intercept of this linear function in the context of this data?
   c. According to the line of best fit, what would be the final English 1 grade for a student with a final Algebra 1 grade of 85?
   d. According to the line of best fit, what would be the final Algebra 1 grade for a student with a final English 1 grade of 75?
   e. Calculate the residuals from the plot above. What do they represent? Are the points with positive residuals located above or below the regression line?
   f. What is the sum of the squared residuals of the linear model that represents the situation described above? Is there any different line that gives a smaller sum? Explain your reasoning.
   g. What is the Least Squares Regression line that models this data? How do you know if this equation is the line of best fit to model the data?
   h. Using technology, compute the correlation coefficient and interpret what it means in the context of the data. (What does the correlation coefficient measure for a linear association?)
   i. Analyze correlation and causation for this data set.

2. The data below shows the number of hours spent studying for the Algebra 1 final exam and the final exam grade of 10 students.

Number of hours	0	1	2	3	4	5	6	7	8	9
Final Exam grade	25	42	53	57	62	73	75	84	87	93

According to the line of best fit, what is the approximate increase in the final exam grade per hour studied for the exam?

PRACTICE TEST

Review All Standards

1. Simplify: (write the answer in exponential form as well as in root form)
   a. $5\sqrt[4]{6} \cdot 2\sqrt[4]{12} \cdot \sqrt[4]{18}$
   b. $\left(x^2 y^{\frac{3}{5}} z^4\right)^{\frac{10}{3}} \cdot \left(x^{\frac{1}{2}} y z^{\frac{2}{3}}\right)^6$
   c. $(8x)^{\frac{1}{3}}\left(x^{\frac{2}{3}}\right)$

2. An athlete runs 16 miles in 3 hours. Use dimensional analysis to convert the athlete's speed to feet per second.

3. On a map, the distance from New York to Washington DC is 1.75 inches. What is the actual distance, knowing that the scale of the map is 1 inch to 150 miles?

4. In one year, the price of the gasoline rose from $3.20 per gallon to $3.60 per gallon. What was the percent increase?

5. Simplify:
$$(3a + 1)(2a - 5) - (a - 3)(2a + 5)$$

6. The width of a *rectangle* is $(4x + 1)$ and the length is $(2x + 5)$. What is the area of the rectangle in terms of x?

7. Factor, using the Greatest Common Factor:
$$12x^2 y + 8x^3 y^3 - 24xy + 16x^2 y^2$$

8. Factor, using the *"Difference of Two Squares"* formula:
$$x^4 - y^4$$

9. Solve the following equation:
$$10b - 4 \cdot (b + 1) = 2 \cdot (3b + 5) - 5 \cdot (2b + 3) - 26$$

10. If $(3m + 2)x - 4x = 2m(6x - 1) + 8m$ is a given equation:
    a. Find the value of x, knowing that $m = 3$.
    b. Find the value of m, knowing that $x = -5$.
    c. Solve the equation for x.
    d. Solve the equation for m.

11. The sum of three numbers is 1032. Find the numbers, knowing that the second number is three times the first number and the third number is 60 more than half of the first number.

12. Graph the equation $y = 2^x$ and find five solutions to this equation. How many solutions does the equation have? How can the solutions be represented on a graph?

13. Graph the equation $2x - 3y + 6 = 0$ and find five solutions to this equation. How many solutions does the equation have? How can the solutions be represented on a graph?

14. Felix has $200 to buy books and movie DVD's. Knowing that a book costs $8 and a movie DVD costs $20, create an equation to represent the relationship between these quantities. Graph this equation and find the viable solutions to the problem. What is the maximum number of books that Felix can buy? What is the maximum number of movie DVD's that Felix can buy?

15. Solve the following system of equations:
$$\begin{cases} 6x - 2y = -14 \\ 7x + y = -13 \end{cases}$$

16. Solve the following system of equations using *graphs* (graph both linear equations and then find the coordinates of the intersection point of the two graphs):
$$\begin{cases} y = 2x - 6 \\ y = -3x + 4 \end{cases}$$

17. Graph the solution to the following systems of linear inequalities:
$$\begin{cases} x + 3y - 6 \geq 0 \\ x \leq 4 \\ y > -2 \end{cases}$$

18. Felix needs to earn at least $850 a week to be able to pay his mortgage and his bills, but he cannot work more than 60 hours per week. Felix is getting paid $20 per hour for mowing lawns and $10 per hour for washing cars.

   a. Create a system of inequalities that represents this situation.
   b. Graph the system and interpret the solution. Find the viable solution in this context.

19. Solve the equation $3^x = -x + 4$ using graphs:

   a. Graph the equations $y = 3^x$ and $y = -x + 4$ in the same coordinate plane.
   b. In how many points do these graphs intersect? What do these points of intersection represent?
   c. Using a table of values for both equations, find approximately the X-coordinate of the point where the graphs intersect.
   d. Using technology, find exactly the coordinates of the point where the graphs of the equations intersect.
   e. What is the solution of the equation $3^x = -x + 4$? What is the solution of the system of equations: $\begin{cases} y = 3^x \\ y = -x + 4 \end{cases}$? Explain why the X-coordinates of the points where the graphs of the equations $y = 3^x$ and $y = -x + 4$ intersect are the solutions of the equation $3^x = -x + 4$.

20. If $f: A \to B$ is a function, and the domain $A = \{1, 2, 3, 4\}$ and $f(x) = x^2 + 2x - 3$, find set B (the range). Complete a table and graph the function f.

21. If $f: A \to B$ is a function, and the range $B = \{1, 2, 3, 4\}$ and $f(x) = 2x - 5$, find set A (the domain). Complete a table and graph the function f.

22. The function $p(t) = 2.7 \cdot (1.08)^t$ gives a city's population $p(t)$ (in millions), where t is the number of years since 1998. According to this function, what was the population of the city in 2010?

23. If $a_n = 2n + 5$ is a sequence, find a_3, a_4, a_{10}.

24. If $x_n = 3 \cdot x_{n-1} + 5$ is a sequence defined recursively and $x_1 = 9$, find x_2, x_5, x_6.

25. The number of bacteria in a culture can be represented by the formula $N_t = 3.7 \cdot N_{t-1}$. In the formula, N_t is the number of bacteria at the end of t minutes, and N_{t-1} is the number of bacteria at the end of $t - 1$ minutes. There are 8,417 bacteria in the culture at the end of 5 minutes. How many bacteria will be in the culture at the end of 9 minutes?

26. Find all the key features (X intercepts, Y intercept, relative minimums, relative maximums, where is the function increasing/decreasing, positive or negative) for the following functions (using maybe the graphing calculator):

    a. $f(x) = 3x + 6$

    b. $f(x) = 2 \cdot e^{4x}$

    c. $f(x) = x^2 + 3x - 4$

27. An arrow is launched from 2 meters above the ground at time $t = 0$. The function that models this situation is given by $h(t) = -\frac{1}{16}t^2 + \frac{7}{8}t + 2$, where t is the time measured in seconds and h is the height above the ground measured in meters.

    a. What is the practical domain for t in this context?

    b. What is the height of the arrow three seconds after it was launched?

    c. What is the maximum value of the function and what does it mean in context?

    d. When is the arrow 3 meters above the ground?

    e. When is the arrow 1 meter above the ground?

    f. What are the intercepts of this function? What do they mean in the context of this problem?

    g. What are the intervals of increase and decrease on the practical domain? What do they mean in the context of this problem?

28. Find the average rate of change of the following function over the specified interval:

$$f(x) = -3x + 5 \text{ ; interval } [1,7]$$

29. Graph the following exponential function, showing intercepts and end behavior:

$$y = -3 \cdot 4^x$$

30. A college's tuition has been increasing 3% each year since 1980. If the tuition cost in 1980 was $7,000, write a function for the amount of the tuition x years after 1980. Find the cost of tuition for this college in 2014.

31. Find the Quadratic function that has the zeros 1 and -4.

32. The annual tuition at a private school since 2004 is modeled by the function $f(x) = 800 \cdot (1.09)^x$, where x is the number of years since 2004.

    a. What was the tuition cost in 2004?

    b. What is the annual percentage of tuition increase?

33. A company earns a weekly profit of P dollars by selling x items according to the function: $P(x) = -x^2 + 25x - 100$

    a. Find the zeros of the function P and interpret these zeros in the context of the problem.

    b. How many items does the company have to sell each week to maximize the profit?

34. Nova Taxi cab company charges a \$3 boarding rate in addition to its meter which is \$2 for every mile. PRITAX cab company charges according to the function $P = 1.5x + 6$, where x is the number of miles traveled.

   a. Compare the charging policies of the two cab companies.

   b. Which cab company has a better price? Explain your answer.

35. Felix's company weekly revenue in dollars is given by the function $R(x) = 2000x - 2x^2$ where x is the number of items produced during a week. Emilia's company weekly revenue is represented in the graph below:

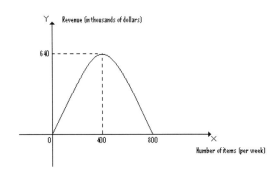

   a. What amount of items will produce the maximum revenue in Felix's company? What amount of items will produce the maximum revenue in Emilia's company?

   b. Compare the evolution of the revenue of the two companies represented by the functions above.

36. The annual tuition at a college A since 2005 is modeled by the function $T(x) = 3000 \cdot (1.03)^x$, where x is the number of years since 2005. The annual tuition at another college B is represented by a graph below:

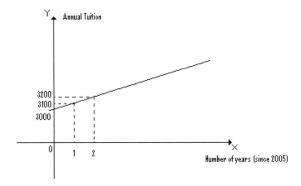

Compare the properties of the two functions and describe the evolution of the tuition cost for the two colleges.

37. Felix wants to invest \$5,000 into a financial plan. Three banks offer the following choices as financial investments:

Bank A: $y = 5000 + 200 \cdot x$

Bank B: $y = 5000 + 50x^2 - 200x$

Bank C: $y = 5000 \cdot (1.02)^x$, where y represents the amount of money after x years .

    a. Which plan is better for Felix if he invests his money for 1 year? How about 5 years? How about 10 years?

    b. Compare the three financial plans using tables or graphs and find which financial plan is the best for different time periods.

38. If $f(x) = 2x - 2$ and $g(x) = x^2 + 2x - 3$ are two functions:

    a. Graph the functions f and g.

    b. Find the expression of the function $h(x) = f(x) + g(x)$.

    c. Sketch the graph of the function h using the graphs of the functions f and g.

    d. Graph the function h, using the explicit form found at point (b) and then compare this graph with the graph from point (c).

39. Write a recursive formula the sequence below (identifying first if it's an arithmetic or geometric sequence)

$$1, 3, 9, 27, 81,...$$

40. For the explicit formula below, write a recursive formula:

$$a_n = 5 \cdot 2^{n-1}$$

41. For the explicit formula below, write a recursive formula:

$$a_n = 5n + 9$$

42. For the recursive formula below, write an explicit formula:

$$a_1 = 19, a_n = a_{n-1} - 4$$

43. For the recursive formula below, write an explicit formula:

$$a_1 = 200, a_n = 0.2 \cdot a_{n-1}$$

44. Felix has \$10,000 to invest in one of the two financial plans offered by a bank. Plan A offers to increase his principal by \$500 each year, while plan B offers to pay 2.5% interest compounded annually.

    a. What plan should Felix choose if he invests his money for 5 years? What if he invests his money for 10 years?

    b. Write a function rule that gives the balance after x years for plan A as well as for plan B.

    c. Compare the two functions, determine whether they represent a linear or an exponential model and find which plan is a better deal for Felix and for what period of time.

45. Which equation represents the graph of $f(x) = 2^{x+1}$ translated 4 units to the left and 3 units up?

46. Construct an exponential function knowing that the graph of this function contains the points $A(3, 4)$ and $B(5, 16)$.

47. Construct a linear function knowing that the graph of this function contains the points $A(-2, 5)$ and $B(-7, -1)$.

48. Bobby wants to invest $5,000 into a financial plan. Three banks offer the following choices as financial investments:

Bank A: $y = 5000 + 200 \cdot x$

Bank B: $y = 5000 + 50x^2 - 200x$

Bank C: $y = 5000 \cdot (1.02)^x$, where y represents the amount of money after x years .

    a. Which plan is better for Bobby if he invests his money for 1 year? How about 5 years? How about 10 years?

    b. Compare the three financial plans using tables or graphs and find which financial plan is the best for different time periods.

49. The annual tuition at a private school since 2001 is modeled by the function $f(x) = 7000 \cdot (1.06)^x$, where x is the number of years since 2004. Interpret the coefficient 7,000 and the base of the exponent 1.06 in the context of this problem.

50. Find the distance between the two given points:

    $G(-3, -7)$ and $H(5, -6)$

51. Write an equation for the line that is *parallel* to the given line and that passes through the given point:

    $y = 4x + 5,$ $A(3,7)$

52. Write an equation for the line that is *perpendicular* to the given line and that passes through the given point:

    $2x - 3y = 9,$ $C(4, -2)$

53. If $M(-3, 4)$ is the midpoint of segment PQ and $P(7, -1)$, what are the coordinates of Q?

54. On a map, Sofia's house is located at $(-9, 1)$ and Tatiana's house is located at $(5, -8)$. What are the coordinates for Diana's house if she lives exactly halfway between Sofia and Tatiana? What is the distance between Sofia's house and Diana's house?

55. The diameter of Planet Earth is 12756.2 kilometers (7926.3 miles).

    a. What is the surface area of Planet Earth?

    b. What is the volume of Planet Earth?

56. You scored an 87 on a test in your Algebra 1 class where the mean was 85 and the standard deviation was 3. Your best friend is in a different Algebra 1 class and scored a 90 where the mean in her class was 88 and the standard deviation was 4. Who had the better score relative to their own class?

57. The data below shows the customer ratings for 2 hotel resorts.

Hotel A: 4.3, 4.1, 3.7, 4.5, 3.2, 4.2, 4.3, 3.8, 4.0, 3.1, 4.9, 3.8, 3.2, 3.6, 4.1

Hotel B: 4.4, 2.5, 3.9, 3.7, 3.8, 2.9, 4.3, 3.6, 4.1, 2.8, 4.9, 3.4, 3.8, 4.2, 3.9

    a. Construct parallel box-plots of the two distributions.

b. Create a histogram for each distribution.

c. Compare center and spread of the two data sets using statistics appropriate to the shape of the distributions.

58. The table below shows the average annual sales (in millions of dollars) at a Shopping Center for the last 10 years.

Year	2002	2003	2004	2005	2006	2007	2008	2009	2010	2011
Sales (in millions)	9.3	9.8	8.7	15.9	7.2	6.9	5.8	7.5	5.4	7.0

a. Find the mean and the standard deviation of the data.

b. In what years were the sales less than one standard deviation from the mean?

c. Find the median and interquartile range for the data.

59. A survey of 18 people asked: "Of SPRING, SUMMER, FALL and WINTER, which is your favorite season of the year?" Each person's occupation was also recorded. The results were as follows:

Season	Occupation
SPRING	ARCHITECT
SUMMER	TEACHER
WINTER	PHARMACIST
SPRING	ARCHITECT
FALL	PHARMACIST
SUMMER	TEACHER
SUMMER	TEACHER
SPRING	ARCHITECT
WINTER	TEACHER
SUMMER	ARCHITECT
FALL	PHARMACIST
SPRING	PHARMACIST
FALL	PHARMACIST
SUMMER	TEACHER
WINTER	ARCHITECT
FALL	PHARMACIST
SUMMER	TEACHER
FALL	ARCHITECT

a. Create a *two-way frequency table* that summarizes these data including *marginal frequencies.* Interpret the *joint frequencies* of this table.

b. Create a *two-way relative frequency table* that summarizes these data which displays *conditional frequencies* for **occupation**. Interpret the *conditional frequencies* of this table.

c. Create a *two-way relative frequency table* that summarizes these data which displays *conditional frequencies* for **season**. Interpret the *conditional frequencies* of this table.

60. The data below shows the final grades for Algebra 1 and English 1 of 15 students.

Algebra 1	82	96	80	84	65	70	78	81	83	68	60	87	90	94	82
English 1	78	82	83	81	68	71	81	98	92	72	65	84	85	88	93

a. Represent the data on a scatter plot and describe how the variables are related using the shape, strength and direction of the scatter plot.

b. Find the linear function that best fits this data. What is the meaning of the slope and the Y-intercept of this linear function in the context of this data?

c. According to the line of best fit, what would be the final English 1 grade for a student with a final Algebra 1 grade of 85?

d. According to the line of best fit, what would be the final Algebra 1 grade for a student with a final English 1 grade of 75?

e. Calculate the residuals from the plot above. What do they represent? Are the points with positive residuals located above or below the regression line?

f. What is the sum of the squared residuals of the linear model that represents the situation described above? Is there any different line that gives a smaller sum? Explain your reasoning.

g. What is the Least Squares Regression line that models this data? How do you know if this equation is the line of best fit to model the data?

h. Using technology, compute the correlation coefficient and interpret what it means in the context of the data. (What does the correlation coefficient measure for a linear association?)

i. Analyze correlation and causation for this data set.

Bibliography

The Common Core State Standards for Mathematics were taken from:
http://maccss.ncdpi.wikispaces.net

Felix Nagy-Lup was born and raised in Cluj-Napoca, Romania.

After graduating from "Babes-Bolyai" University with a degree in Mathematics he received his Master's degree in Algebra. In 2008 the North Carolina Council of Teachers of Mathematics awarded him as their choice for the MATH TEACHER OF THE YEAR for Harnett County Schools based on his students' success at the "End of Course" state exams.

"Successful Mathematics" is his second workbook. He previously published a similar book for his students in Romania.

Mr. Nagy-Lup is currently a mathematics teacher at Harnett Central High School in North Carolina. He lives in Lillington, North Carolina with his wife and his daughter.

VISIT: successfulmathematics.blogspot.com

ORDER INFORMATION:

     E-MAIL: felixnagylup@gmail.com

     PHONE: 910-496-6726